T0142132

Advances in Experimental Medicine and Biology

Clinical and Experimental Biomedicine

Volume 1251

Series Editor
Mieczyslaw Pokorski
Opole Medical School
Opole, Poland

More information about this series at http://www.springer.com/series/16003

Mieczyslaw Pokorski
Editor

Trends in Biomedical Research

Editor
Mieczyslaw Pokorski
Opole Medical School
Opole, Poland

ISSN 0065-2598 ISSN 2214-8019 (electronic)
Advances in Experimental Medicine and Biology
ISSN 2523-3769 ISSN 2523-3777 (electronic)
Clinical and Experimental Biomedicine
ISBN 978-3-030-41221-0 ISBN 978-3-030-41219-7 (eBook)
https://doi.org/10.1007/978-3-030-41219-7

This Springer imprint is published by the registered company Springer Nature Switzerland AG.
The registered company address is: Gewerbestrasse 11, 6330 Cham, Switzerland

Contents

Targeted Drug Delivery from Titanium Implants: A Review of Challenges and Approaches 1
Anwesha Barik and Nishant Chakravorty

Reliability of Magnetic Resonance Tractography in Predicting Early Clinical Improvements in Patients with Diffuse Axonal Injury Grade III 19
Sunil Munakomi, Deepak Poudel, and Sangam Shrestha

Integrated Thermal Rehabilitation: A New Therapeutic Approach for Disabilities 29
Giovanni Barassi, Esteban Obrero-Gaitan, Giuseppe Irace, Matteo Crudeli, Giovanni Campobasso, Francesco Palano, Leonardo Trivisano, and Vito Piazzolla

Cat Allergy as a Source Intensification of Atopic Dermatitis in Adult Patients 39
Andrzej Kazimierz Jaworek, Krystyna Szafraniec, Magdalena Jaworek, Zbigniew Doniec, Adam Zalewski, Ryszard Kurzawa, Anna Wojas–Pelc, and Mieczyslaw Pokorski

Clinical Course and Outcome of Community-Acquired Bacterial Meningitis in Cancer Patients 49
Marcin Paciorek, Agnieszka Bednarska, Dominika Krogulec, Michał Makowiecki, Justyna D. Kowalska, Dominik Bursa, Anna Świderska, Joanna Puła, Joanna Raczyńska, Agata Skrzat-Klapaczyńska, Marek Radkowski, Urszula Demkow, Tomasz Laskus, and Andrzej Horban

Influence of Coping Strategy on Perception of Anxiety and Depression in Patients with Non-small Cell Lung Cancer 57
Beata Jankowska-Polańska, Jacek Polański, Mariusz Chabowski, Joanna Rosińczuk, and Grzegorz Mazur

Inflammatory Markers During Continuous High Cutoff Hemodialysis in Patients with Septic Shock and Acute Kidney Injury 71
Grzegorz Kade, Sławomir Literacki, Agnieszka Rzeszotarska, Stanisław Niemczyk, and Arkadiusz Lubas

Body Composition and Biochemical Markers of Nutrition in Non-dialysis-Dependent Chronic Kidney Disease Patients 81
Aleksandra Rymarz, Maria Zajbt, Anna Jeznach-Steinhagen, Agnieszka Woźniak-Kosek, and Stanisław Niemczyk

Biocompatibility of Hemodialysis 91
Małgorzata Gomółka, Longin Niemczyk, Katarzyna Szamotulska, Magdalena Mossakowska, Jerzy Smoszna, Aleksandra Rymarz, Leszek Pączek, and Stanisław Niemczyk

Decreasing Vaccination Coverage Against Hepatitis B and Tuberculosis in Newborns 99
Aneta Nitsch-Osuch, Beata Pawlus, Maria Pawlak, and Ernest Kuchar

Virological and Epidemiological Situation in the Influenza Epidemic Seasons 2016/2017 and 2017/2018 in Poland 107
E. Hallmann-Szelińska, K. Łuniewska, K. Szymański, D. Kowalczyk, R. Sałamatin, A. Masny, and L. B. Brydak

Epidemic Influenza Seasons from 2008 to 2018 in Poland: A Focused Review of Virological Characteristics 115
Sainjargal Byambasuren, Iwona Paradowska-Stankiewicz, and Lidia B. Brydak

Adv Exp Med Biol - Clinical and Experimental Biomedicine (2020) 8: 1–17
https://doi.org/10.1007/5584_2019_447

Published online: 26 November 2019

Targeted Drug Delivery from Titanium Implants: A Review of Challenges and Approaches

Anwesha Barik and Nishant Chakravorty

Abstract

Titanium implants are considered the gold standard of treatment for dental and orthopedic applications. Biocompatibility, low elasticity, and corrosion resistance are some of the key properties of these metallic implants. Nonetheless, a long-term clinical failure of implants may occur due to inadequate osseointegration. Poor osseointegration induces mobility, inflammation, increased bone resorption, and osteolysis; hence, it may result in painful revision surgeries. Topographical modifications, improvement in hydrophilicity, and the development of controlled-release drug-loading systems have shown to improve cellular adhesion, proliferation, and differentiation. Surface modifications, along with drug coating, undoubtedly demonstrate better osseointegration, especially in challenged degenerative conditions, such as osteoporosis, osteoarthritis, and osteogenesis imperfecta. Anabolic bone-acting drugs, such as parathyroid hormone peptides, simvastatin, prostaglandin-EP4-receptor antagonist, vitamin D, strontium ranelate, and anti-catabolic bone-acting drugs, such as calcitonin, bisphosphonates, and selective estrogen receptor modulators, expedite the process of osseointegration. In addition, various proteins, peptides, and growth factors may accessorize the idea of localized therapy. Loading these substances on modified titanium surfaces is achieved commonly by mechanisms such as direct coating, adsorption, and incorporating in biodegradable polymers. The primary approach toward the optimum drug loading is a critical trade-off between factors preventing release of a drug immediately and those allowing slow and sustained release. Recent advances broaden the understanding of the efficacy of adsorption, hydrogel coating, and electrospinning layer-by-layer coating facilitated by differential charge on metallic surface. This review discusses the existing approaches and challenges for the development of stable and sustained drug delivery systems on titanium implants, which would promote faster and superior osseointegration.

Keywords

Biocompatibility · Biodegradable polymers · Drug coating · Drug delivery · Metallic surface · Osseointegration · Targeted drugs · Titanium implants

A. Barik and N. Chakravorty (✉)
School of Medical Science and Technology, Indian Institute of Technology Kharagpur, Kharagpur, Paschim Medinipur, West Bengal, India
e-mail: nishant@smst.iitkgp.ac.in

1 Introduction

Implantology, as a specialized branch of orthopedics and dentistry, has been recognized by clinician and researchers globally for quite some time.

With the discovery of the phenomenon of "osseointegration" in the 1950–1960s, titanium became the material of choice for most clinicians in the field of dental and orthopedic implants (Brånemark et al. 2001). While implant surgeries often result in positive outcomes, premature failure is noted in about 10% of the cases (Kurtz et al. 2005). Such rare undesirable consequences arise from complications which eventually fail to provide enough mechanical support to diseased skeletal structures. Some of the primary reasons that require attention include insufficient tissue-metal integration, bacterial invasion, and foreign body reaction. In fact, these issues may become critical even after many years after successful osseointegration. Colonization of bacteria borne by saliva on the exposed surfaces of implants is the harbinger of distress and results in peri-implantitis, i.e., a progressive loss of surrounding bone (Mombelli et al. 1987; Rams et al. 1984). Inadequate bone and soft tissue quality at the site of insertion also attributes to implant failure (Javed et al. 2013). Therefore, osseointegration still remains an important challenge in implantology research.

Researchers and clinicians have always been of the opinion that supplemental treatment with drugs may help in improving the bone binding ability of implants. Although systemic administration of drugs along with implant placement is a conventional approach, side effects, such as systemic toxicity, short-term benefits, and patient inconvenience, limit their use. In contrast to these, localized therapy is generally devoid of these limitations (Anselmo and Mitragotri 2014; Park et al. 2014). Lately, focus has been placed on bone-specific delivery of active pharmaceutical and biological agents. Antibiotic-loaded polymethyl methacrylate (PMMA) bone cement was first reported in the 1970s as an adjunct therapy for bone implants to reduce implant-associated infection (Bistolfi et al. 2011; Masri et al. 1998; Chapman and Hadley 1976; Holm and Vejlsgaard 1976; Marks et al. 1976; Buchholz and Engelbrecht 1970). A global commercialization of such products has eventually happened. However, these products received several criticisms, such as nonbiodegradability of the cement material used, erratic drug release due to permanent binding of most of the drug within the cement, possible drug degradation during the in situ exothermic process of self-curing, and others (Tobin 2017). Other FDA-approved biodegradable polymers were tried to incorporate drug molecules, which may be exemplified by poly lactic acid (PLA) or poly lactic-co-glycolic acid (PLGA). However, high temperature reactions that occur during the chemical bonding phase of most materials have discouraged their further use. Lessons learned from the past have guided us toward the use of modern implants coated with drug molecules using novel processes that avoid the aforementioned concerns. Simple, low temperature, reproducible processes, which can be optimized to guide extended release patterns of drugs, have become the present day methods of choice. Other ideal characteristics of such devices include the absence or minimal systemic toxicity by targeting at the organ, tissue, or cellular level, which allows the use of a lower drug dose. Additional care needs to be taken during the fabrication process, to retain the original mechanical properties of an implant.

Standardization of therapeutic delivery of drugs is a challenging process for multiple reasons. Desirable drug release needs optimization of several factors, such as delivery vehicle, type of drug carrier, thickness, composition, focal tissue microenvironment, and physicochemical parameters of a drug. Moreover, standardization is dependent on the clinical requirements which, in turn, depend on the severity of a disease. A number of approaches toward a biological coating on titanium implants are reported and discussed in this review. The discussion highlights the current trending technology in coated orthopedic and dental devices, which is considered to mitigate the lack of localized bonding of implants to bone. To this end, collating a comprehensive list of features inclusive of minimum toxicity, enhanced device performance, or tunable release kinetics is a quite challenging task. New methods such as tethering of antimicrobial peptide molecules, incorporation of bone-specific growth factors and proteins, or bio-mimicking next-generation coatings are still gaining importance as evidenced from several in vitro and in vivo studies. Nonetheless, supporting clinical evaluation studies that would provide a better understanding and validation of these approaches are yet to come.

2 Osseointegration of Titanium Alloy Implants

Titanium and its alloys are recognized as highly promising materials for dental and orthopedic applications. Commercially, pure titanium is generally used for dental implants due to limited mechanical properties whereas alloys are suitable for prolonged high load bearing applications such as hip joint replacements, bone screws, or plates (Shah et al. 2016). The implants need to be fused completely with the surrounding tissues and provide the platform for cellular adhesion, proliferation, and maturation. Osseointegration is considered to be established, provided that the peri-implant bone is capable to resist surrounding shear forces and to maintain the maximum gap of 50 μm between implant and tissue to prevent forming fibrous capsule (Bloebaum et al. 1994). Bone formation around machined implants occurs through new bone deposition on existing bone tissue (distance osteogenesis) and direct osteoblastic activity on microroughened implants via contact osteogenesis (Yamaki et al. 2012; Amor et al. 2011; Davies 2003). Overall, this process usually starts by bone apposition in trabecular bone and then toward formation of compact bone, which can be histologically observed within 1 week of insertion, and the growth continues lifelong thereafter (Slaets et al. 2007). Osseointegration predominantly depends on numerous biological, physical, chemical, thermal, and other factors. Mechanical properties of titanium suggest that high dielectric constant of titanium oxide layer (naturally produced/deposited) assists in bonding tissue with metal surface.

Despite the success of titanium as a bone implant, several clinical trials have reported failures upon long-term evaluations, and lack of osseointegration seems the culprit in most cases (Santos et al. 2002; Esposito et al. 1999). Over the years, several modifications have been attempted to improve bone-binding ability of implant surfaces. Topographical modifications by physical or chemical surface treatment to induce porosity or addition of a biocompatible coating are useful as shown in in vitro and in vivo experiments (Salou et al. 2015; Donos et al. 2011; Bjursten et al. 2010; Li et al. 2004; Klokkevold et al. 1997) Such modified surfaces assist the bone cells to grow on the surface and down through the interconnected pores, tubes, and channels (Mandracci et al. 2016). The patterns induce roughness to increase the surface area of the modified implants compared to the smooth polished ones (Le Guehennec et al. 2007). Present-day concepts concerning implant instability emphasize that topographies in the micro-to-nano ranges improve implant fixation (Ferraris et al. 2015; Davies et al. 2013). Imparting surface roughness, especially in a range of 1–10 μm, shows positive effects in terms of biomolecular interaction between bone and implant (Damiati et al. 2018; Liang et al. 2017; Gittens et al. 2014). Other studies have reported different ranges of pores to be conducive to cellular anchorage, estrogenic differentiation, and bone ingrowth along with vascular system development, including some studies where pore diameters up to about few hundred microns seems favorable (Vasconcellos et al. 2008; Otsuki et al. 2006). Nano-scale (defined as less than 100 nm) modifications of titanium implants are recently gaining interest owing to their ability to not only change physical but also the chemical interaction patterns capable to produce effects on the interface biology. Nano-roughening, reportedly, increases the contact area of titanium with human bone "grains", which are usually 20–80 nm long and 2–3 nm in diameter for osteoid minerals (Staruch et al. 2016). This is believed to increase cell adhesion energy and thereby leads to better protein adsorption, cellular proliferation, and differentiation (Xie et al. 2017). Some studies have failed to differentiate the osteogenic activity between micro- and nano-scale modifications, except the tendency for upregulation of genes on nano-structured surfaces (Barik et al. 2017). The primary objective of the surface treatment is to reduce the micromovement of implants during the post-implantation period, so that stability is achieved at the earliest. Thereafter, full osseointegration may only be assured.

3 Use of Drugs: Chemical Ways for Improvement

Combinatorial therapy is always considered to be a better choice of treatment, especially in clinical situations prone to develop complications, or where there is a need for prophylactic action to prevent failure of therapy. Biofilm formation and lack of osseointegration are the two main factors impeding the long-term success of titanium implants. Use of antibiotics to prevent biofilm formation and several anti-resorptive and anabolic agents to ensure implant-bone binding are undoubtedly the mainstream treatment modalities being considered since years.

Choice of antibiotics to combat the risk of susceptible infections primarily depends on the type of infective organism. *Staphylococcus aureus*, *Pseudomonas aeruginosa*, *Staphylococcus epidermidis,* and *Escherichia coli* have been identified as the most common organisms involved in implant-associated infections and osteomyelitis (Ribeiro et al. 2012). A broad range of antibiotics have been tested in vitro and in vivo. These include cephalosporins, aminoglycosides, quinolones, carbapenems, and glycopeptides (Romano et al. 2015). Considering the fact that systemic administration of antibiotics prior to surgery has been found to be ineffective in up to 4% of cases of total hip and knee arthroplasties (Stigter et al. 2004), a need for localized drug therapy has been emerging. This proposition has paved the way to develop drug-incorporated implants. Surface tethering of implants to prevent bacterial adhesion and colonization has been widely investigated by either surface coating with antibiotics or by incorporating antimicrobial molecules through covalent attachments. Imparting bioactivity in this way disrupts the microbes on attachment to the surface, resulting in inhibition of biofilm formation (Ketonis et al. 2012). Slow and sustained elution of immobilized drug should be ensured to reach the effective concentration at a constant rate for a minimum of 4 weeks in addition to peak release for initial 4–7 days (Pan et al. 2018). Achieving the effective minimum inhibitory concentration (MIC) of the antibiotic at the site is equally important, aside from sustained release of a drug, since lower drug concentrations might induce microbial resistance and higher ones might produce toxic reactions.

Bone growth around the grafted implants is a continuous cyclic process balanced by homeostatic activity between osteoclasts and osteoblasts. Osteoclast-mediated bone resorption needs to be balanced at one end of this feedback loop. Bisphosphonates (alendronate, zoledronate, or risedronate) are the commonest agents belonging to anti-resorptive drugs used to mediate increased osteoblastic activity. The presence of carbon as a bridge between phosphate ions in the chemical structure helps the drug to bind the divalent cations, especially Ca^{2+} present in hydroxyapatite in the site for active bone remodeling (Qayoom et al. 2018). Several clinical, preclinical, and in vitro studies have evaluated the efficacy of estrogen, denosumab, and selective estrogen receptor modulators (SERM) such as raloxifene, although mostly without using implants (Faverani et al. 2018; Steffi et al. 2017; Januario et al. 2001; Tıraş et al. 2000). These molecules chemically mimic the biological activity of estrogen – a natural bone protector. They bind to the same ligand-binding domain present in estrogen receptors (ER), producing a conformational change with some minor differences. As a result, they promote further binding of other co-regulatory proteins that activate the transcriptional process acting on specific consensus ER response elements in the target gene promoter regions, the way estrogen naturally does (Migliaccio et al. 2007). This anti-resorptive therapy exerts the overall protective effect on the bone by reducing the activity, life span, and a number of osteoclasts, which stabilize trabecular microarchitecture, preserves bone mass, and bone mineral density, decreasing the risk of fractures.

Anabolic drugs are another group of pharmacological agents that play a positive role in improving bone mineral density, aiding both cortical and trabecular bone formation. Preclinical and clinical studies suggest that they may increase overall bone strength by positively modulating bone remodeling (Brandi 2012). Teriparatide, parathyroid hormone, or anti-sclerostin antibodies are some examples of anabolic agents (Shibamoto et al. 2018; Suzuki et al. 2017, 2018; Lai et al. 2017; Virdi et al. 2015). Strontium ranelate is a compound that has garnered a lot of attention in treatment of osteoporosis, owing to its unique properties of possessing both anti-catabolic and anabolic effects. The presence of both anti-resorptive and pro-bone forming abilities restores bone formation by balancing bone metabolism (Meunier et al. 2004). Furthermore, possession of similar mechanical and chemical properties akin to calcium allows easy incorporation of strontium into the mineral phase of bone, pointing to its effectiveness in treating osteoporosis (Kyllonen et al. 2015; Li et al. 2012a; Maïmoun et al. 2010). There is a multitude of pharmacological tools that appears effective in inducing bone formation and inhibiting bone resorption. Surprisingly, very few of those drugs have so far been tried for coating on implants. Bisphosphonates are one of the most common classes of drugs that have been investigated for this kind of localized release.

4 Implant Coating Strategy: Essential Points

Immobilization technology on titanium implants is still in its stage of infancy. Despite the plurality of research in the area, experiments are still restricted to in vitro cell culture and there are few preclinical studies. Systemic administration, i.e., oral, subcutaneous, or intramuscular is approved as the most suitable mode of treatment. Localized drug administration is less promising due to variabilities in drug loading. Additionally, local pharmacokinetics in such type of drug delivery still needs to be defined. Quantitative determination of released drugs is complicated as in vivo adverse interactions, such as enzymatic degradation, colloidal aggregation, or unwanted ligand binding with cell receptors, decrease the bioavailability of a drug at the peri-implant region (Pokrowiecki 2018). Sustained drug release at the site of action is much more advantageous in the long term, provided that the drug dosing could be suitably controlled. Overall, pharmacokinetics and pharmacodynamics of small molecules can be improved by controlling the rate of drug dissolution and diffusion through maintaining a steady rate of degradation of a carrier. Coating on metallic implants, like titanium, may be idealized by featuring a few characteristics (Zemtsova et al. 2018; Tobin 2017; Santos et al. 2014) which are represented in the following flowchart (Fig. 1).

5 Types of Implant Coating: What We Know

A number of biologically active materials can act as potential candidates for coating implant surfaces with growth factors and drugs. Various ways of coatings are inspired by inorganic (hydroxyapatite) and organic (extracellular matrix and proteins) components of the bone or by a combination of components, which truly resembles bone (de Jonge et al. 2008). Drug properties such as solubility, potency, and desired site of action and clearance rate each influences a suitable selection of a desired drug delivery mode (Damiati et al. 2018). Additionally, choice of the mode determines the drug loading capacity, duration of release, and the most appropriate route for administration. Features of a drug delivery mode, such as size, surface charge and hydrophobicity, shape, flexibility, and inclusion of targeting moieties, affect performance and its bio-distribution. The resultant drug release from any carrier is determined by a complex interaction between the drug properties, polymer characteristics, and in vivo conditions.

5.1 Films

Two-dimensional film formation on titanium implants by coating is the simplest strategy for localized drug delivery. Physical adsorption of a

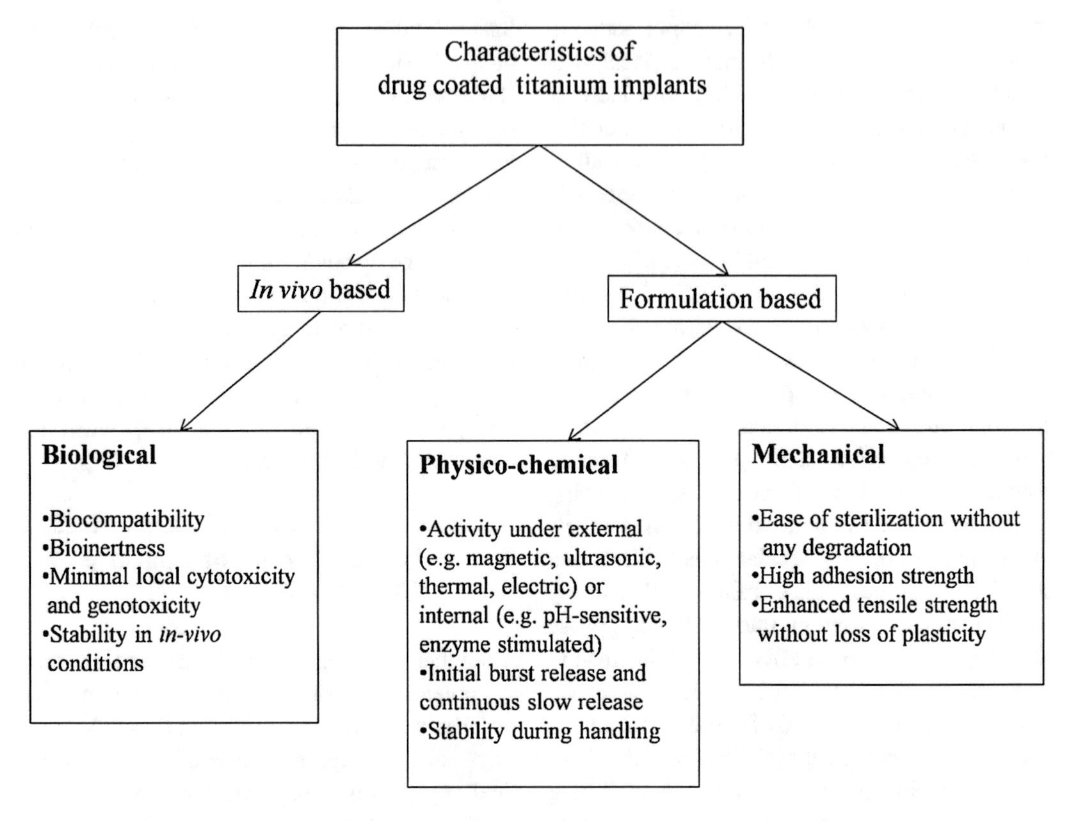

Fig. 1 Ideal characteristics of coatings on metallic implants

drug-incorporated polymer solution on the surface by means of dipping, spray coating, or drop casting techniques are in common practice (Corobea et al. 2015; Cortizo et al. 2012). Despite simplicity, reproducibility among batches is a major concern with this method. Lack of uniformity of single coatings and inability to standardize the amount of a drug to be loaded into the thin film make the technique questionable for appropriate implementation. In addition, control of various parameters such as chemical composition, thickness, or architecture also needs to be taken care of (Boudou et al. 2010). Polyelectrolyte multilayer films (PEM), which is conformal electrostatically bonded thin film by alternate dip/spray coating of oppositely charged polyelectrolytes, have been evaluated to get a reproducible uniform coating on titanium substrates. Loading of BMP-2, an osteo-conductive agent, on PEM appears quite uniform and also provides a uniform release rate in bioactivity assays as well as in vivo studies (Guillot et al. 2016). This self-assembly process starts with the dipping of an oppositely charged polyelectrolyte solution on a charged substrate to form a monolayer by absorption. The step is repeated with intermittent washing after each coating to remove weakly bound or unbound species. A variety of parameters may complicate the otherwise simple and straightforward process. The pH, concentration, ionic strength of the polyelectrolyte solution, and the time of deposition need to be highly optimized to standardize the morphology, thickness, and biological features of the film (Shi et al. 2017). Occasionally, an initiator is needed to impart a stable charge to the surface in an attempt to further improve the adhesion of a film. Poly-L-lysine is one such initiator compound in common practice (Shu et al. 2011). Another application of this layer-by-layer (LBL) technique has been reported by Qian et al. (2014), where the authors used an alternate coating with heparin and chitosan and they loaded an anti-osteoporotic drug (HU-308) by simple adsorption on the film. Coated implants showed new bone formation and improved osseointegration at an early-to-middle stage of placement.

5.2 Hydrogels

Hydrogels are best defined as hydrophilic interpenetrating 3D networks of one, two, or more polymers formed by physical and chemical crosslinking of the participating molecules. Hydrogels inherently have high swelling properties when in contact with water or any other type of body fluid, while maintaining their 3D shape, mechanical, and elastic properties (Sosnik and Seremeta 2017). Porosity, an attractive feature for their widespread use in tissue engineering, can be tuned by the extent of crosslinking reaction and the interstitial spaces may be used to load therapeutic compounds. These interlinked pores help diffuse drug molecules within tissues or surrounding cells and help pass nutrients and factors through the gel to provide a platform for cellular growth (Concheiro and Alvarez-Lorenzo 2013; Anumolu et al. 2009). Synthetic polymers, such as polyethylene oxide (PEO), polyacrylic acid (PAA), and polyvinyl alcohol (PVA), and natural ones, such as silk fibroin, dextran, alginate, chitosan, and gelatin, are commonly used to formulate hydrogels that provide a solubilizing environment to incorporate hydrophilic drugs (Wei et al. 2014). The amount of a drug released is very critical during the design of a hydrogel-based local drug delivery system. Antibiotic-laden hyaluronic acid-based hydrogels impair bone formation only up to certain concentration of a drug (Boot et al. 2017). A fast resorbable hydrogel composed of covalently linked hyaluronian and poly-D,L-lactide inhibits bacterial colonization by immediate release of antibiotics (Drago et al. 2014). Sustained long-term drug release is most desirable for regenerative purposes. Recently, a complex localized drug delivery approach was attempted by Li et al. (2017), where the drug-incorporating four-armed thiolated PEG-hydrogel was used to retard the initial burst release. Crosslinked starch was used to further reduce the possibility of fast swelling kinetics of hydrogel. Polydopamine-assisted surface modification enhances the adhesion stability of a hydrogel film. Several coating procedures and the release of drug molecules from films and hydrogels are presented in Fig. 2.

5.3 Microparticles

This submicron particulate drug delivery system offers a number of advantages including cell-specific drug delivery following endocytosis. Such systems are particularly useful to sustain a drug level at a localized region while minimizing systemic toxicity (Kohane 2007). Controlled release kinetics of drugs from microparticles depends on the rate of water penetration in the carrier and subsequent hydrolytic degradation of the polymer. A hollow structure of spheres allows the maximum entrapment of a drug amount to be loaded. Hydrophobic polymers, such as poly-lactic-co-glycolic acid (PLGA) or laboratory-designed co-polymers having tunable hydrophilic and hydrophobic properties, e.g., thioether-containing ω-hydroxy acid (TEHA), block copolymers with poly-caprolactone, or TEHA with poly ethylene glycol, are usually the ones selected for the formation of microparticles loaded with hydrophobic drugs (Nerantzaki et al. 2018). Immobilization or attachment of microparticles on the titanium implants may require physical or chemical treatment. For instance, Dawes et al. (2010) have thermally immobilized dexamethasone-loaded microspheres on electrolytically oxidized titanium surfaces. A similar method has been employed to render the metal hydrophilic, with subsequent loading of microspheres performed by polyethyleneimine treatment (Son et al. 2013). Vacuum drying and attachment at room temperature are some of the other methods that have been attempted for bonding of microspheres to metal surface (Xiao et al. 2014). One of the challenges with such delivery systems is to retard the fast release of hydrophilic drugs due to its high

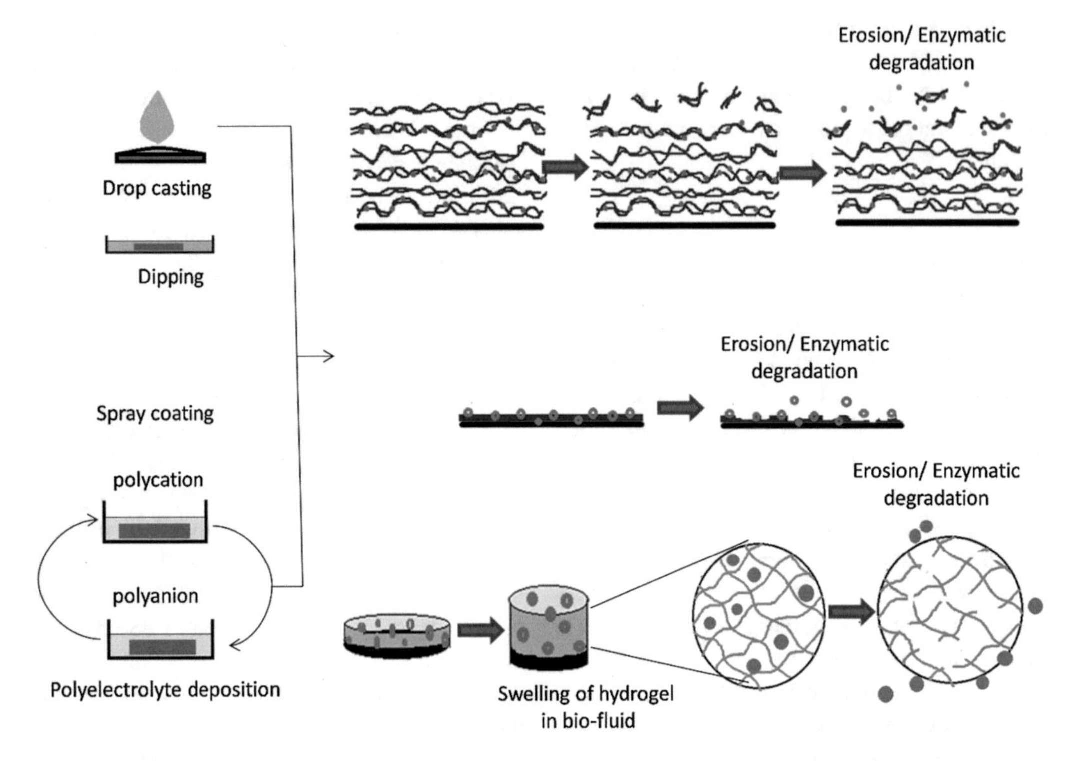

Fig. 2 Coating of films/hydrogels on titanium surfaces

aqueous solubility. Chitosan and sodium alginate are some of the hydrophilic polymers which have been tried to develop the drug-loaded microspheres, and these agents may be entrapped into the porous titanium surface as reported by Wang et al. (2015).

5.4 Nanoparticles

Nanoparticles (typically ranged 1–100 nm) are considered as ideal vehicles for drug delivery that allow fast accumulation of drugs in the targeted regions. Nanoparticles as drug delivery systems are usually more advantageous than microparticles due to a greater proportion of drug entrapment and the ability to cross most of the biological barriers (Kohane 2007). Considering a high surface-to-volume ratio of nanoparticles, such systems may occasionally exhibit a greater toxicity, owing to the high encapsulation and release of a drug at the site. The nano-scale of particles also is capable of triggering immunological responses when they come in contact with host cells, and hence limits the increase in their use for drug delivery (Moreno-Vega et al. 2012). Despite the limitation, nanoparticle coating on implants imparts osteo-conductivity due to a high surface area, aside from the other advantages as outlined above. Chitosan nanoparticles for delivery of bone morphogenetic protein 2 (BMP-2) and ciprofloxacin have been studied successfully by coating on titanium implants, using spray coating and drop casting in two different studies (Poth et al. 2015; De Giglio et al. 2012). In the former study, a degree of acetylation of chitosan was tailored for fast biodegradation by lysozyme, which expectedly helped in a dose-dependent ectopic bone growth in vivo. In the later study, ionic gelation principle was used to form β-cyclodextrin-based inclusion complexes between ciprofloxacin molecules itself and between chitosan and ciprofloxacin molecule while preparing nanoparticles, with the intention to obtain maximum release within first 5 days post-implantation. In both cases, nanoparticles were confirmed to be biocompatible, highly encapsulated, and had a narrow size distribution. Nanoparticle-based prolonged release has been found feasible in some studies. A combination of other surface modifications prior to nanoparticle loading, such as electrophoretic deposition of ciprofloxacin-loaded hydroxyapatite nanoparticles on acid-etched titanium, is another successful strategy for prolonged release as described by Geuli et al. (2017). Ibuprofen-loaded mesoporous silica nanoparticles and chitosan were electrolytically co-deposited to form another novel drug release system which, after changing the pH, eventually formed interpenetrating chitosan hydrogel network containing nanoparticles (Zhao et al. 2014). This system shows zero-order drug release strongly influenced by pH and both electrical stimuli. The coating of microparticles/ nanoparticles on titanium surfaces has been briefly presented in Fig. 3.

5.5 Electrospun Nanofibers

Nanofibers are characterized by non-woven highly porous structure with high permeability, stability, surface area, and ease of functionalization. These fibers are produced by electrospinning technique. This simple setup operates with a high voltage power supply that allows the spinneret to spin polymer solutions and allow formation of fibers that can be collected on a grounded metal collector at ambient conditions (Al-Enizi et al. 2018). Extracellular matrix of most of the human organs and tissues, for instance, bone and skin which need prostheses for regeneration, consists of organized hierarchical fibrous interwoven structures. Coating of such prostheses with fiber mats resembling the natural microporous structure enables the support of cellular anchorage within the pores and provides a sufficient area for cell growth (Kenry and Lim 2017; Khadka and Haynie 2012). Such fibrous coatings have been shown to inhibit implant failure by modifying the mismatched interface between hard metal and soft tissue (Al-Enizi et al. 2018). Production of the electrospun nanofibers is generally performed using polymers or materials, such as PLGA, which have similar biocompatibility features to those used for the other delivery systems as discussed above (Ashbaugh et al. 2016; Zhang et al. 2014; Li et al. 2012b). Drug-loaded nanofibers can be produced either by blending the drug with polymer prior to electrospinning or by adsorption and immobilization of factors on the fibers afterward. Notably, incorporation of hydrophilic drugs in hydrophobic polymers during blending is not possible due to solubility issues. To overcome this problem and to attain sustained release, investigators have successfully tried co-axial electrospinning while entrapping the drug in a hydrophilic core surrounded by a hydrophobic sheath (Song et al. 2017). Simple hydrophilic polymer combinations, such as chitosan and polyethylene oxide, have also been used for loading of vancomycin. Another alternative is pretreatment of titanium implants with coating of chitosan and gelatin for better attachment of nanofibers (Sadri et al. 2017). A combination of surface treatment with polymer-assisted drug delivery has always shown better results when compared to either one. These two separate features synergistically help in osseointegration and anti-infective properties, depending on the drug to be delivered. In vivo results show better bone growth on titanium surfaces with microgrooves filled with drug-loaded nanofibers (Khandaker et al. 2017). The process is limited by the concurrent effects of multiple variables acting together, such as the voltage applied, flow rate of solution, diameter of needle, distance between needle and the collector, or the polymer solution properties, such as viscosity, molecular weight, dielectric constant, and surface tension. Control of the porosity is a real research challenge since it depends on the diameter of the formed fibers, which, in turn, is depended on multiple factors as mentioned above. Other challenges include lack of infiltration of cells within smaller pores that may be formed due to the smaller diameter of fibers, leading to inapplicability of the coated device in

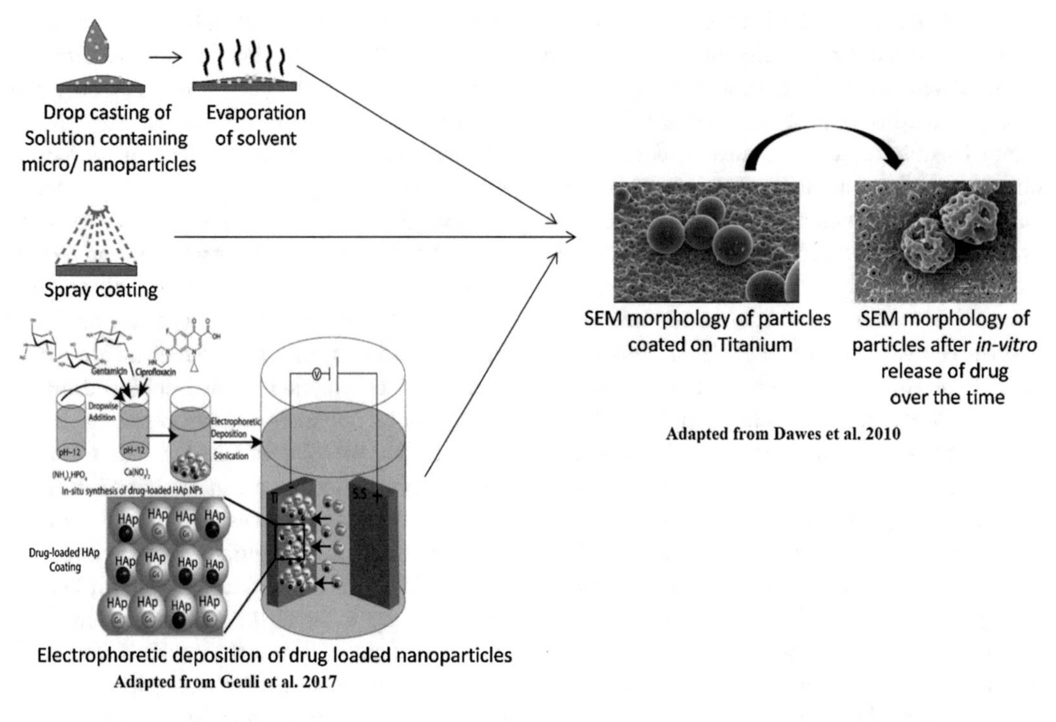

Fig. 3 Coating of micro-/nanoparticles on titanium surfaces

tissue engineering (Dahlin et al. 2011). The formation of electrospun nanofibers and encapsulation of drugs is represented in Fig. 4.

5.6 Titanium Nanotubes

Physical adsorption of drugs on the TiO_2 nanotube arrays to enable controlled release at a site of administration is considered an innovative smart technology and has been suggested for applications in dentistry and orthopedics. Excellent in vitro and in vivo biocompatibility and high surface area due to the presence of nano-porous organization provide a good template for drug loading, which is better than those with other strategies, and also support the spread of bone cells inside the nanotopographies. Low immunogenicity due to the complete absence of any foreign materials glorifies the great potential of nanotubes to be introduced into bone tissue engineering (Ainslie et al. 2009). Self-assembled porous nanostructures are produced upon the establishment of equilibrium between the reactions in the electrochemical anodization process. The oxide formation on the surface after dissolution of the metal occurs parallel with the formation of vertically aligned porous nanostructure on the oxide layer dissolution (Wang et al. 2016). A variety of parameters involved in the anodization, such as reaction temperature, anodization voltage, or reaction time, controls the features of nanotubes, such as the diameter, length, aspect ratio, and volume. These features are highly correlated with the drug release rate which mostly follows diffusion-limited processes, defined by Fick's first law (Azhang et al. 2015). Hydrophilic drug loading in nanotubes usually follows a simple process of physical adsorption of the aqueous solution. For controlled diffusion of drugs, introduction of a biocompatible polymeric coating, such as PLGA or chitosan, over the nanotube arrays may be beneficial to tune and extend the burst and sustained release as necessary. The coating also favors the cellular attachment and proliferation (Wang et al. 2017; Gulati et al. 2012). Apart from topographical modification and biodegradable polymer

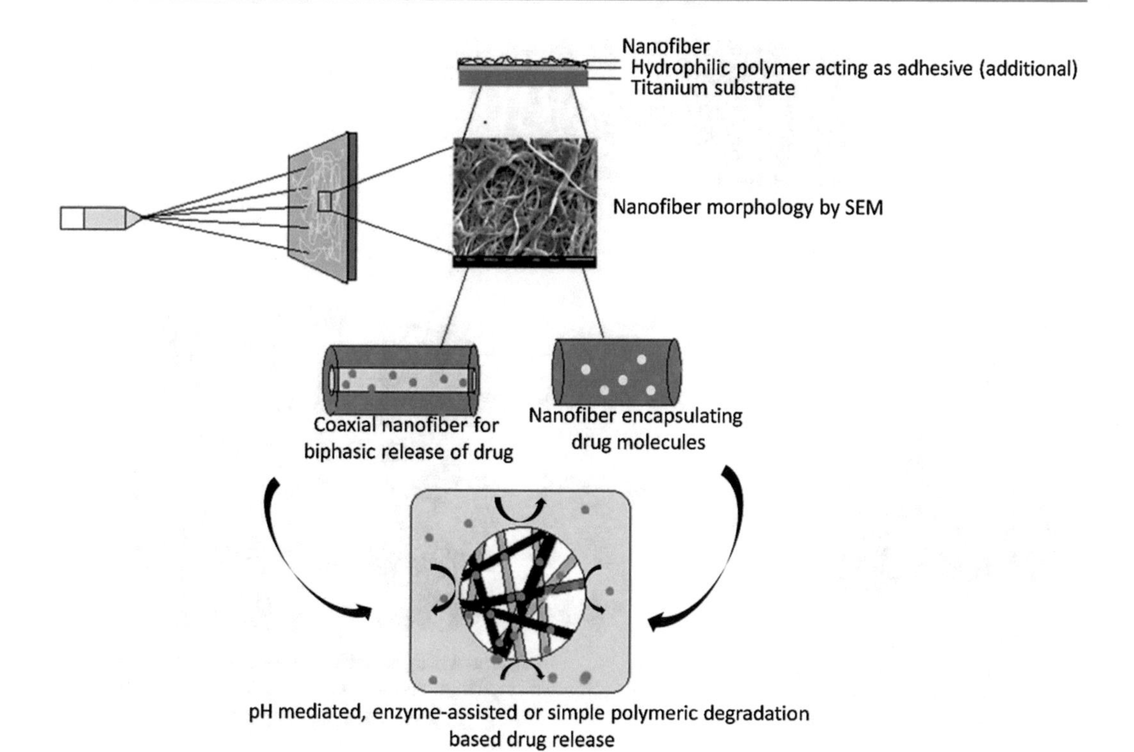

Fig. 4 Coating of electrospun nanofibers on titanium implants and encapsulation of drug molecules. *SEM,* scanning electron microscopy

coatings, drug loading in, and release from, nanotubes can be controlled by introducing some other carriers, such as micelles or nanoparticles, especially to encapsulate hydrophobic drugs and to combine hydrophobic and hydrophilic drugs (Aw et al. 2012). Drug delivery by titanium nanotubes is summarized in Fig. 5.

6 Limitations in the Current Technology and Future Perspectives: Bridging the Gap

While drug-releasing implants are currently well established and quite standardized for the intervention in cardiovascular diseases and diabetes, the development of bone-specific localized drug delivery has not been fully implemented into clinical practice. Therapeutic benefits need to be confirmed through a rigorous path of translational research, from in vitro to in vivo to clinical trials, to reduce the risk of clinical failure. Investigations in experimental animals do not guarantee success in human application (Anselmo and Mitragotri 2014). Apart from economic constraints and limitations in clinical feasibility, each of the approaches of drug coating on titanium implants provides some benefits, but also has inherent drawbacks. Keeping the upcoming technologies in mind, customized 3D printing of drug-loaded titanium implants might be a realistic solution in the near future. The advent of 3D printing technology in the area of customizable orthopedics is expected to rise. The 3D printouts, free from complex scaffolds and metal mono-blocks, are completely devoid of stiffness mismatch at bone-implant interface due to reprogramming of the stress shielding pattern by variation of the porosity. This interconnected porous network aids in rapid bone in-growth, if the shape and porosity can be controlled by rapid prototyping method, which makes it a suitable candidate for orthopedic applications (Lopez-Heredia et al. 2008). Elastic modulus also needs to be adjusted

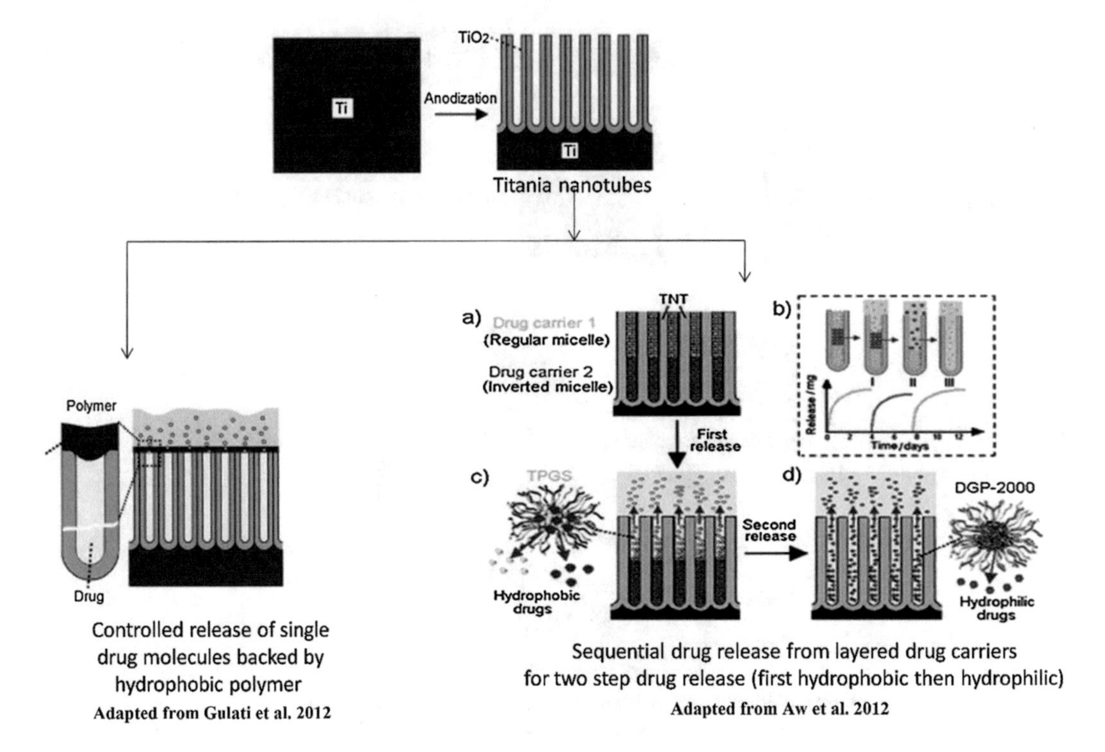

Fig. 5 Localized drug delivery through titanium nanotubes

according to that of the natural tissue (Heinl et al. 2008). The production of 3D printed porous scaffold incorporated with drugs has already been used for a case of surgery in tuberculous spinal infection where poly-D,L-lactase and nano-hydroxyapatite were blended with anti-tubercular drugs (Dong et al. 2014). Spinal tuberculosis-induced bone defect has been successfully reconstructed by this scaffold, ensuring spinal stabilization.

Controlled drug release from titanium implant surfaces has been recently made possible by a selective laser melting process (Hassanin et al. 2018). It is hypothesized that the drug release kinetics through the drug-loaded internalized compartments and microchannels can be predicted and modified as desired. These topographies are optimized to achieve a minimum deviation in microchannel dimension and surface roughness at 1–2 μm, depending upon the laser power. Despite investigations pointing to better endurance of laser-deposited titanium implants, clinical application of such implants is still limited (Chan et al. 2013). If implemented, it holds an excellent opportunity for implant-mediated targeted drug delivery system.

Smart biosensing technique can also be utilized to switch on and off the release of a drug on demand depending on the severity and stage of a defect as well as progression of improvement. Sirivisoot et al. (2011) have developed a smart anodized nano-tubular titanium implant where anti-inflammatory and antibiotic molecules are electro-deposited on poly-pyrrole, a polymer that can be electrochemically synthesized on conductive materials. These carbon nanotubes act as biosensing electrodes to detect the type of cellular attachment or tissue formation around the implant, based on differences in electrical conductivity between inflammatory cells, bone cells, and bacteria. Those authors have shown the feasibility of delivering a drug of choice, once a particular biological event is determined. Titanium implant-attached biochemical sensing to assess blood indices may also become a way of noninvasive disease monitoring (Li and Lu 2015), a technique that can be associated with the drug delivery platform.

7 Conclusion

This review summarized different forms of titanium implants, based on drug delivery modes which have been successively developed and modified over recent years. The most challenging clinical situation arising post-implantation is a peri-implant infection and lack of osseointegration. Biologics and drug molecules have a pivotal role to ensure the stability and long-activity of implants, whether administered locally or systemically. The experimental phase of research into these aspects continues to deliver stable biologic coating on orthopedic implants, in which the type, carrier, method, thickness, stability, and in vivo degradation constitute some of the important concerns. The predicted mathematical relationship between mechanical degradation of coating and diffusion kinetics of drug molecules stored in the reservoir should be mimicked in vivo. An effective commercial technique should be designed to advance industrial production. Apart from these, all the stakeholders, especially surgeons, should be clearly aware of the proper indications for storage and use of titanium implants to avoid any misuse or unintended failure.

Conflicts of Interest The authors declare that they have no conflicts of interest in relation to this article.

Ethical Approval This review article does not contain any studies with human participants or animals performed by any of the authors.

References

Ainslie KM, Tao SL, Popat KC, Daniels H, Hardev V, Grimes CA, Desai TA (2009) In vitro inflammatory response of nanostructured titania, silicon oxide, and polycaprolactone. J Biomed Mater Res A 91 (3):647–655

Al-Enizi A, Zagho M, Elzatahry A (2018) Polymer-based electrospun nanofibers for biomedical applications. Nano 8(4):259

Amor N, Geris L, Vander Sloten J, Van Oosterwyck H (2011) Computational modelling of biomaterial surface interactions with blood platelets and osteoblastic cells for the prediction of contact osteogenesis. Acta Biomater 7(2):779–790

Anselmo AC, Mitragotri S (2014) An overview of clinical and commercial impact of drug delivery systems. J Control Release 190:15–28

Anumolu SS, Singh Y, Gao D, Stein S, Sinko PJ (2009) Design and evaluation of novel fast forming pilocarpine-loaded ocular hydrogels for sustained pharmacological response. J Control Release 137 (2):152–159

Ashbaugh AG, Jiang X, Zheng J, Tsai AS, Kim WS, Thompson JM, Thompson JM, Miller RJ, Shahbazian JH, Wang Y, Dillen CA, Ordonez AA, Chang YS, Jain SK, Jones LC, Sterling RS, Mao HQ, Miller LS (2016) Polymeric nanofiber coating with tunable combinatorial antibiotic delivery prevents biofilm-associated infection in vivo. Proc Natl Acad Sci U S A 113(45): E6919–E6928

Aw MS, Addai-Mensah J, Losic D (2012) A multi-drug delivery system with sequential release using titania nanotube arrays. Chem Commun (Camb) 48 (27):3348–3350

Azhang H, Suman SR, Christos T, Mathew TM, Cortino S, Alexander LY, Tolou S (2015) Fabrication of drug eluting implants: study of drug release mechanism from titanium dioxide nanotubes. J Phys D Appl Phys 48(27):275401

Barik A, Banerjee S, Dhara S, Chakravorty N (2017) A reductionist approach to extract robust molecular markers from microarray data series – isolating markers to track osseointegration. J Biomed Inform 68:104–111

Bistolfi A, Massazza G, Verné E, Massè A, Deledda D, Ferraris S, Miola M, Galetto F, Crova M (2011) Antibiotic-loaded cement in orthopedic surgery: a review. ISRN Orthop 2011:290851

Bjursten LM, Rasmusson L, Oh S, Smith GC, Brammer KS, Jin S (2010) Titanium dioxide nanotubes enhance bone bonding in vivo. J Biomed Mater Res A 92A (3):1218–1224

Bloebaum RD, Bachus KN, Momberger NG, Hofmann AA (1994) Mineral apposition rates of human cancellous bone at the interface of porous coated implants. J Biomed Mater Res 28(5):537–544

Boot W, Gawlitta D, Nikkels PGJ, Pouran B, van Rijen MHP, Dhert WJA, Vogely HC (2017) Hyaluronic acid-based hydrogel coating does not affect bone apposition at the implant surface in a rabbit model. Clin Orthop Relat Res 475(7):1911–1919

Boudou T, Crouzier T, Ren K, Blin G, Picart C (2010) Multiple functionalities of polyelectrolyte multilayer films: new biomedical applications. Adv Mater 22 (4):441–467

Brandi ML (2012) Drugs for bone healing. Expert Opin Investig Drugs 21(8):1169–1176

Brånemark R, Brånemark PI, Rydevik B, Myers RR (2001) Osseointegration in skeletal reconstruction and rehabilitation: a review. J Rehabil Res Dev 38:175–181

Buchholz HW, Engelbrecht H (1970) Depot effects of various antibiotics mixed with palacos resins. Chirurg 41(11):511–515

Chan KS, Koike M, Mason RL, Okabe T (2013) Fatigue life of titanium alloys fabricated by additive layer manufacturing techniques for dental implants. Metall Mater Trans 44(2):1010–1022

Chapman MW, Hadley WK (1976) The effect of polymethylmethacrylate and antibiotic combinations on bacterial viability. An in vitro and preliminary in vivo study. Bone Joint Surg Am 58(1):76–81

Concheiro A, Alvarez-Lorenzo C (2013) Chemically cross-linked and grafted cyclodextrin hydrogels: from nanostructures to drug-eluting medical devices. Adv Drug Deliv Rev 65(9):1188–1203

Corobea MS, Albu MG, Ion R, Cimpean A, Miculescu F, Antoniac IV, Raditoiu V, Sirbu I, Stoenescu M, Voicu SI, Ghica MV (2015) Modification of titanium surface with collagen and doxycycline as a new approach in dental implants. J Adhes Sci Technol 29 (23):2537–2550

Cortizo MC, Oberti TG, Cortizo MS, Cortizo AM, Fernández Lorenzo de Mele MA (2012) Chlorhexidine delivery system from titanium/polybenzyl acrylate coating: evaluation of cytotoxicity and early bacterial adhesion. J Dent 40(4):329–337

Dahlin RL, Kasper FK, Mikos AG (2011) Polymeric nanofibers in tissue engineering. Tissue Eng Part B Rev 17(5):349–364

Damiati L, Eales MG, Nobbs AH, Su B, Tsimbouri PM, Salmeron-Sanchez M, Dalby MJ (2018) Impact of surface topography and coating on osteogenesis and bacterial attachment on titanium implants. J Tissue Eng 9:2041731418790694

Davies JE (2003) Understanding peri-implant endosseous healing. J Dent Educ 67(8):932–949

Davies JE, Ajami E, Moineddin R, Mendes VV (2013) The roles of different scale ranges of surface implant topography on the stability of the bone/implant interface. Biomaterials 34(14):3535–3546

Dawes GJ, Fratila-Apachitei LE, Necula BS, Apachitei I, Witkamp GJ, Duszczyk J (2010) Release of PLGA-encapsulated dexamethasone from microsphere loaded porous surfaces. J Mater Sci Mater Med 21(1):215–221

De Giglio E, Trapani A, Cafagna D, Ferretti C, Iatta R, Cometa S, Ceci C, Romanelli A, Mattioli-Belmonte M (2012) Cirpofloxacin-loaded chitosan nanoparticles as titanium coatings: a valuable strategy to prevent implant associated infections. Nano Biomed Eng 4 (4):162–168

de Jonge LT, Leeuwenburgh SC, Wolke JG, Jansen JA (2008) Organic-inorganic surface modifications for titanium implant surfaces. Pharm Res 25 (10):2357–2369

Dong J, Zhang S, Liu H, Li X, Liu Y, Du Y (2014) Novel alternative therapy for spinal tuberculosis during surgery: reconstructing with anti-tuberculosis bioactivity implants. Expert Opin Drug Deliv 11(3):299–305

Donos N, Hamlet S, Lang NP, Salvi GE, Huynh-Ba G, Bosshardt DD, Ivanovski S (2011) Gene expression profile of osseointegration of a hydrophilic compared with a hydrophobic microrough implant surface. Clin Oral Implants Res 22(4):365–372

Drago L, Boot W, Dimas K, Malizos K, Hänsch GM, Stuyck J, Gawlitta D, Romanò CL (2014) Does implant coating with antibacterial-loaded hydrogel reduce bacterial colonization and biofilm formation in vitro? Clin Orthop Relat Res 472(11):3311–3323

Esposito M, Lausmaa J, Hirsch JM, Thomsen P (1999) Surface analysis of failed oral titanium implants. J Biomed Mater Res 48(4):59–568

Faverani LP, Polo TOB, Ramalho-Ferreira G, Momesso GAC, Hassumi JS, Rossi AC, Freire AR, Prado FB, Luvizuto ER, Gruber R, Okamoto R (2018) Raloxifene but not alendronate can compensate the impaired osseointegration in osteoporotic rats. Clin Oral Investig 22(1):255–265

Ferraris S, Bobbio A, Miola M, Spriano S (2015) Micro- and nano-textured, hydrophilic and bioactive titanium dental implants. Surf Coat Technol 276:374–383

Geuli O, Metoki N, Zada T, Reches M, Eliaz N, Mandler D (2017) Synthesis, coating, and drug-release of hydroxyapatite nanoparticles loaded with antibiotics. J Mater Chem B 5(38):7819–7830

Gittens RA, Olivares-Navarrete R, Schwartz Z, Boyan BD (2014) Implant osseointegration and the role of microroughness and nanostructures: lessons for spine implants. Acta Biomater 10(8):3363–3371

Guillot R, Pignot-Paintrand I, Lavaud J, Decambron A, Bourgeois E, Josserand V, Logeart-Avramoglou D, Viguier E, Picart C (2016) Assessment of a polyelectrolyte multilayer film coating loaded with BMP-2 on titanium and PEEK implants in the rabbit femoral condyle. Acta Biomater 36:310–322

Gulati K, Ramakrishnan S, Aw MS, Atkins GJ, Findlay DM, Losic D (2012) Biocompatible polymer coating of titania nanotube arrays for improved drug elution and osteoblast adhesion. Acta Biomater 8(1):449–456

Hassanin H, Finet L, Cox SC, Jamshidi P, Grover LM, Shepherd DET, Addison O, Attallah MM (2018) Tailoring selective laser melting process for titanium drug-delivering implants with releasing micro-channels. Addit Manuf 20:144–155

Heinl P, Müller L, Körner C, Singer RF, Müller FA (2008) Cellular Ti-6Al-4V structures with interconnected macro porosity for bone implants fabricated by selective electron beam melting. Acta Biomater 4 (5):1536–1544

Holm NJ, Vejlsgaard R (1976) The in vitro elution of gentamicin sulphate from methylmethacrylate bone cement. A comparative study. Acta Orthop Scand 47 (2):44–148

Januario AL, Sallum EA, de Toledo S, Sallum AW, Nociti JF Jr (2001) Effect of calcitonin on bone formation around titanium implant. A histometric study in rabbits. Braz Dent J 12(3):158–162

Javed F, Ahmed HB, Crespi R, Romanos GE (2013) Role of primary stability for successful osseointegration of dental implants: factors of influence and evaluation. Interv Med Appl Sci 5(4):162–167

Kenry, Lim CT (2017) Nanofiber technology: current status and emerging developments. Prog Polym Sci 70:1–17

Ketonis C, Parvizi J, Jones LC (2012) Evolving strategies to prevent implant-associated infections. J Am Acad Orthop Surg 20(7):478–480

Khadka D, Haynie DT (2012) Protein- and peptide-based electrospun nanofibers in medical biomaterials. Nanomedicine 8(8):1242–1262

Khandaker M, Riahinezhad S, Williams WR, Wolf R (2017) Microgroove and collagen-poly(epsilon – caprolactone) nanofiber mesh coating improves the mechanical stability and osseointegration of titanium implants. Nanomaterials (Basel) 7(6):145

Klokkevold PR, Nishimura RD, Adachi M, Caputo A (1997) Osseointegration enhanced by chemical etching of the titanium surface. A torque removal study in the rabbit. Clin Oral Implants Res 8(6):442–447

Kohane DS (2007) Microparticles and nanoparticles for drug delivery. Biotechnol Bioeng 96(2):203–209

Kurtz S, Mowat F, Ong K, Chan N, Lau E, Halpern M (2005) Prevalence of primary and revision total hip and knee arthroplasty in the United States from 1990 through 2002. J Bone Joint Surg Am 87(7):1487–1497

Kyllonen L, D'Este M, Alini M, Eglin D (2015) Local drug delivery for enhancing fracture healing in osteoporotic bone. Acta Biomater 11:412–434

Lai K, Xi Y, Miao X, Jiang Z, Wang Y, Wang H, Yang G (2017) PTH coatings on titanium surfaces improved osteogenic integration by increasing expression levels of BMP-2/Runx2/Osterix. RSC Adv 7 (89):56256–56265

Le Guehennec L, Soueidan A, Layrolle P, Amouriq Y (2007) Surface treatments of titanium dental implants for rapid osseointegration. Dent Mater 23(7):844–854

Li YJ, Lu CC (2015) A novel scheme and evaluations on a long-term and continuous biosensor platform integrated with a dental implant fixture and its prosthetic abutment. Sensors (Basel) 15(10):24961–24976

Li LH, Kong YM, Kim HW, Kim YW, Kim HE, Heo SJ, Koak JY (2004) Improved biological performance of Ti implants due to surface modification by micro-arc oxidation. Biomaterials 25(14):2867–2875

Li Y, Li X, Song G, Chen K, Yin G, Hu J (2012a) Effects of strontium ranelate on osseointegration of titanium implant in osteoporotic rats. Clin Oral Implants Res 23 (9):1038–1044

Li LL, Wang LM, Xu Y, Lv LX (2012b) Preparation of gentamicin-loaded electrospun coating on titanium implants and a study of their properties in vitro. Arch Orthop Trauma Surg 132(6):897–903

Li D, Lv P, Fan L, Huang Y, Yang F, Mei X, Wu D (2017) The immobilization of antibiotic-loaded polymeric coatings on osteoarticular Ti implants for the prevention of bone infections. Biomater Sci 5(11):2337–2346

Liang J, Xu S, Shen M, Cheng B, Li Y, Liu X, Qin D, Bellare A, Kong L (2017) Osteogenic activity of titanium surfaces with hierarchical micro−/nano-structures obtained by hydrofluoric acid treatment. Int J Nanomedicine 12:1317–1328

Lopez-Heredia MA, Goyenvalle E, Aguado E, Pilet P, Leroux C, Dorget M, Weis P, Layrolle P (2008) Bone growth in rapid prototyped porous titanium implants. J Biomed Mater Res A 85(3):664–673

Maïmoun L, Brennan TC, Badoud I, Dubois-Ferriere V, Rizzoli R, Ammann P (2010) Strontium ranelate improves implant osseointegration. Bone 46 (5):1436–1441

Mandracci P, Mussano F, Rivolo P, Carossa S (2016) Surface treatments and functional coatings for biocompatibility improvement and bacterial adhesion reduction in dental implantology. Coatings 6(1):7

Marks KE, Nelson CL, Lautenschlager EP (1976) Antibiotic-impregnated acrylic bone cement. J Bone Joint Surg Am 58(3):358–364

Masri BA, Duncan CP, Beauchamp CP (1998) Long-term elution of antibiotics from bone-cement: an in vivo study using the prosthesis of antibiotic-loaded acrylic cement (PROSTALAC) system. J Arthroplasty 13 (3):331–338

Meunier PJ, Roux C, Seeman E, Ortolani S, Badurski JE, Spector TD, Cannata J, Balogh A, Lemmel EM, Pors-Nielsen S, Rizzoli R, Genant HK, Reginster JY (2004) The effects of strontium ranelate on the risk of vertebral fracture in women with postmenopausal osteoporosis. N Engl J Med 350(5):459–468

Migliaccio S, Brama M, Spera G (2007) The differential effects of bisphosphonates, SERMS (selective estrogen receptor modulators), and parathyroid hormone on bone remodeling in osteoporosis. Clin Interv Aging 2 (1):55–64

Mombelli A, van Oosten MA, Schürch E Jr, Lang NP (1987) The microbiota associated with successful or failing osseointegrated titanium implants. Oral Microbiol Immunol 2(4):145–151

Moreno-Vega AI, Gómez-Quintero T, Nuñez-Anita RE, Acosta-Torres LS, Castaño V (2012) Polymeric and ceramic nanoparticles in biomedical applications. J Nanotechnol 2012:10

Nerantzaki M, Skoufa E, Adam KV, Nanaki S, Avgeropoulos A, Kostoglou M, Bikiaris D (2018) Amphiphilic block copolymer microspheres derived from castor oil, poly(ε-carpolactone), and poly(ethylene glycol): preparation, characterization and application in naltrexone drug delivery. Materials (Basel) 11 (10). https://doi.org/10.3390/ma11101996

Otsuki B, Takemoto M, Fujibayashi S, Neo M, Kokubo T, Nakamura T (2006) Pore throat size and connectivity determine bone and tissue ingrowth into porous implants: three imensional micro-CT based structural analyses of porous bioactive titanium implants. Biomaterials 27(35):5892–5900

Pan C, Zhou Z, Yu X (2018) Coatings as the useful drug delivery system for the prevention of implant-related infections. J Orthop Surg Res 13(1):220

Park YS, Cho JY, Lee SJ, Hwang CI (2014) Modified titanium implant as a gateway to the human body: the implant mediated drug delivery system. Biomed Res Int 2014:801358

Pokrowiecki R (2018) The paradigm shift for drug delivery systems for oral and maxillofacial implants. Drug Deliv 25(1):1504–1515

Poth N, Seiffart V, Gross G, Menzel H, Dempwolf W (2015) Biodegradable chitosan nanoparticle coatings on titanium for the delivery of BMP-2. Biomol Ther 5(1):3–19

Qayoom I, Raina DB, Širka A, Tarasevičius Š, Tägil M, Kumar A, Lidgren L (2018) Anabolic and antiresorptive actions of locally delivered bisphosphonates for bone repair. Bone Joint Res 7 (10):548–560

Qian X, Qing F, Jun O, Hong S (2014) Construction of drug-loaded titanium implants via layer-by-layer electrostatic self-assembly. Hua Xi Kou Qiang Yi Xue Za Zhi 32(6):537–541. (Article in Chinese)

Rams TE, Roberts TW, Tatum H, Keyes PH (1984) The subgingival microbial flora associated with human dental implants. J Prosthet Dent 51(4):529–534

Ribeiro M, Monteiro FJ, Ferraz MP (2012) Infection of orthopedic implants with emphasis on bacterial adhesion process and techniques used in studying bacterial-material interactions. Biomatter 2(4):176–194

Romano CL, Scarponi S, Gallazzi E, Romano D, Drago L (2015) Antibacterial coating of implants in orthopaedics and trauma: a classification proposal in an evolving panorama. J Orthop Surg Res 10:157

Sadri M, Pashmfroosh N, Samadieh S (2017) Implants modified with polymeric nanofibers coating containing the antibiotic vancomycin. Nanomed Res J 2 (4):208–215

Salou L, Hoornaert A, Louarn G, Layrolle P (2015) Enhanced osseointegration of titanium implants with nanostructured surfaces: an experimental study in rabbits. Acta Biomater 11:494–502

Santos MCLG, Campos MIG, Line SRP (2002) Early dental implant failure: a review of the literature. Braz J Oral Sci 1:103–111

Santos A, Sinn Aw M, Bariana M, Kumeria T, Wang Y, Losic D (2014) Drug-releasing implants: current progress, challenges and perspectives. J Mater Chem B 2 (37):6157–6182

Shah FA, Trobos M, Thomsen P, Palmquist A (2016) Commercially pure titanium (cp-Ti) versus titanium alloy (Ti6Al4V) materials as bone anchored implants – is one truly better than the other? Mater Sci Eng C 62:960–966

Shi Q, Qian Z, Liu D, Liu H (2017) Surface modification of dental titanium implant by layer-by-layer electrostatic self-assembly. Front Physiol 8:574–574

Shibamoto A, Ogawa T, Duyck J, Vandamme K, Naert I, Sasaki K (2018) Effect of high frequency loading and parathyroid hormone administration on peri-implant bone healing and osseointegration. Int J Oral Sci 10 (1):6

Shu Y, Ou G, Wang L, Zou J, Li Q (2011) Surface modification of titanium with heparin-chitosan multilayers via layer-by-layer self-assembly technique. J Nanomater 2011:8

Sirivisoot S, Pareta R, Webster TJ (2011) Electrically controlled drug release from nanostructured polypyrrole coated on titanium. Nanotechnology 22 (8):085101

Slaets E, Carmeliet G, Naert I, Duyck J (2007) Early trabecular bone healing around titanium implants: a histologic study in rabbits. J Periodontol 78 (3):510–517

Son JS, Choi YA, Park EK, Kwon TY, Kim KH, Lee KB (2013) Drug delivery from hydroxyapatite-coated titanium surfaces using biodegradable particle carriers. J Biomed Mater Res B Appl Biomater 101B(2):247–257

Song W, Seta J, Chen L, Bergum C, Zhou Z, Kanneganti P, Kast RE, Auner GW, Shen M, Markel DC, Ren W, Yu X (2017) Doxycycline-loaded coaxial nanofiber coating of titanium implants enhances osseointegration and inhibits Staphylococcus aureus infection. Biomed Mater 12(4):045008

Sosnik A, Seremeta K (2017) Polymeric hydrogels as technology platform for drug delivery applications. Gels 3(3):25

Staruch R, Griffin MF, Butler P (2016) Nanoscale surface modifications of orthopaedic implants: state of the art and perspectives. Open Orthop J 10:920–938

Steffi C, Shi Z, Kong CH, Wang W (2017) In vitro findings of titanium functionalized with estradiol via polydopamine adlayer. J Funct Biomater 8(4):45

Stigter M, Bezemer J, de Groot K, Layrolle P (2004) Incorporation of different antibiotics into carbonated hydroxyapatite coatings on titanium implants, release and antibiotic efficacy. J Control Release 99 (1):127–137

Suzuki T, Ryu K, Kojima K, Saito S, Nagaoka H, Tokuhashi Y (2017) Teriparatide treatment improved loosening of cementless total knee arthroplasty: a case report. J Orthop Case Rep 7(1):32–35

Suzuki T, Sukezaki F, Shibuki T, Toyoshima Y, Nagai T, Inagaki K (2018) Teriparatide administration increases periprosthetic bone mineral density after total knee arthroplasty: a prospective study. J Arthroplast 33 (1):79–85

Tıraş MB, Noyan V, Yıldız A, Yıldırım M, Daya S (2000) Effects of alendronate and hormone replacement therapy, alone or in combination, on bone mass in postmenopausal women with osteoporosis: a prospective, randomized study. Hum Reprod 15(10):2087–2092

Tobin EJ (2017) Recent coating developments for combination devices in orthopedic and dental applications: a literature review. Adv Drug Deliv Rev 112:88–100

Vasconcellos LMR, Oliveira MV, Graça MLA, Vasconcellos LGO, Cairo CAA, Carvalho YR (2008) Design of dental implants, influence on the osteogenesis and fixation. J Mater Sci Mater Med 19 (8):2851–2857

Virdi AS, Irish J, Sena K, Liu M, Ke H, McNulty M, Sumner DR (2015) Sclerostin antibody treatment improves implant fixation in a model of severe osteoporosis. J Bone Joint Surg Am 97(2):133–140

Wang D, Liu Q, Xiao D, Guo T, Ma Y, Duan K, Wang J, Lu X, Feng B, Weng J (2015) Microparticle entrapment for drug release from porous-surfaced bone implants. J Microencapsul 32(5):443–449

Wang Q, Huang JY, Li HQ, Chen Z, Zhao AZ, Wang Y, Zhang KQ, Sun HT, Al-Deyab SS, Lai YK (2016) TiO2 nanotube platforms for smart drug delivery: a review. Int J Nanomedicine 11:4819–4834

Wang T, Weng Z, Liu X, Yeung KWK, Pan H, Wu S (2017) Controlled release and biocompatibility of polymer/titania nanotube array system on titanium implants. Bioact Mater 2(1):44–50

Wei Q, Becherer T, Angioletti-Uberti S, Dzubiella J, Wischke C, Neffe AT, Lendlein A, Ballauff M, Haag R (2014) Protein interactions with polymer coatings and biomaterials. Angew Chem Int Ed Engl 53 (31):8004–8031

Xiao D, Liu Q, Wang D, Xie T, Guo T, Duan K, Weng J (2014) Room temperature attachment of PLGA microspheres to titanium surfaces for implant-based drug release. Appl Surf Sci 309:112–118

Xie Y, Li J, Yu ZM, Wei Q (2017) Nano modified SLA process for titanium implants. Mater Lett 186:38–41

Yamaki K, Kataoka Y, Ohtsuka F, Miyazaki T (2012) Micro-CT evaluation of in vivo osteogenesis at implants processed by wire-type electric discharge machining. Dent Mater J 31(3):427–432

Zemtsova EG, Arbenin AY, Valiev RZ, Smirnov VM (2018) Improvement of the mechanical and biomedical properties of implants via the production of nanocomposite based on nanostructured titanium matrix and bioactive nanocoating. In: Anisimov K et al (eds) Proceedings of the scientific-practical conference 'Research and Development – 2016'. Springer, Cham

Zhang L, Yan J, Yin Z, Tang C, Guo Y, Li D, Wei B, Xu Y, Gu Q, Wang L (2014) Electrospun vancomycin-loaded coating on titanium implants for the prevention of implant-associated infections. Int J Nanomedicine 9:3027–3036

Zhao P, Liu H, Deng H, Xiao L, Qin C, Du Y, Shi X (2014) A study of chitosan hydrogel with embedded mesoporous silica nanoparticles loaded by ibuprofen as a dual stimuli-responsive drug release system for surface coating of titanium implants. Colloids Surf B Biointerfaces 123:657–663

Adv Exp Med Biol - Clinical and Experimental Biomedicine (2020) 8: 19–28
https://doi.org/10.1007/5584_2019_445

Published online: 3 December 2019

Reliability of Magnetic Resonance Tractography in Predicting Early Clinical Improvements in Patients with Diffuse Axonal Injury Grade III

Sunil Munakomi, Deepak Poudel, and Sangam Shrestha

Abstract

Diffuse axonal injury (DAI) grade III forms a distinct subset of traumatic brain injury wherein it is difficult to predict the outcome and the time taken for early recovery in terms of sustained eye opening and standing with minimal assistance. This study seeks to determine differences in the fractional anisotropy (FI) and diffusion-weighted image (DWI) values obtained from the seeds placed at an appropriate region of interest (ROI) within the magnetic resonance (MR) tractography of the brainstem of brain-injured patients. We found that differences in the DWI values along the corticospinal tract were associated with the days required for early recovery. Moreover, dysautonomia was an independent variable governing a delayed recovery in these patients. The lesions posterior to the corticospinal tract in the brainstem conferred increased odds for the subsequent development of dysautonomia. We conclude that MR tractography, in addition to depicting the anatomical integrity of the concerned tracts, has the potential of becoming a surrogate clinical imaging marker for effectively predicting days for early recovery among patients with DAI grade III.

Keywords

Axonal injury · Brain · Corticospinal tracts · Dysautonomia · Magnetic resonance · Tractography

1 Introduction

Diffuse axonal injury (DAI) grade III encompasses a distinct traumatic brain entity that has a profound effect on the neurological outcome of patients owing to its effects on the neural fiber tracts that modulate vital functions, such as motion via corticospinal tracts or arousal via the reticular activating system (Gennarelli 1987). Such patients are most often in poor neurological status at the time of presentation, which necessitates intensive care support. A pivotal concern is that of the time required for the patient recovery. The only manner the family members relate to the improvement of their loved ones is either through noticing spontaneous eye opening or seeing them standing with support. These changes provide strength and believing in the

S. Munakomi (✉)
Department of Neurosurgery, Nobel Medical College and Teaching Hospital, Biratnagar, Nepal
e-mail: sunilmunakomi@gmail.com

D. Poudel
Department of General Anesthesia, Nobel Medical College and Teaching Hospital, Biratnagar, Nepal

S. Shrestha
Department of Pediatrics, Koshi Zonal Hospital, Biratnagar, Nepal

possibility of further recovery (van Eijck et al. 2018b). The treating doctors are compelled to counsel the families concerning the ongoing supporting management and the observation for signs of recovery in patients with DAI, without any definitive armamentarium to predict days to recovery. Moreover, in low-income nations, these factors are a major determinant of family decision to continue an extended hospital management. Early signs of recovery will motivate patients to continue treatment, hoping for improvement despite financial constraints.

In this study, we set out to evaluate the possible role of diffusion tensor imaging (DTI) in quantifying and predicting days to recovery among patients with DAI. Recovery prediction would also help to properly dichotomize patients, thereby allowing for a suitable allocation of health care resources. Computerized tomography (CT) is useful only in ruling out other obvious intracranial pathologies or showing scattered brain tissue contusions, which may be conducive to making a provisional diagnosis of DAI. On the other hand, magnetic resonance sequences consisting of gradient recall echo, susceptibility-weighted imaging, and diffusion-weighted imaging (DWI) are now considered the modalities of choice in a discrete localization and anatomical grading of DAI (Ma et al. 2016; Ashwal et al. 2006). DTI or neural fiber tractography helps in providing a complete anatomical tracing of selected nerve fibers in the region of interest, which is of help in differentiating any abrupt cutoff, compression, or wasting of neural tracts and thus has a prognostic value (Mac Donald et al. 2007; Inglese et al. 2005). The model we propose in this study may provide a predictive knowledge concerning the time needed for potential recovery in patients with DAI.

2 Methods

This study comprised the patients with a provisional diagnosis of DAI, who were admitted to the intensive care unit of the Department of Neurosurgery at the Nobel Medical College and Teaching Hospital in Biratnagar, Nepal from January 2018 to January 2019. Eventually, there were 17 patients included with DAI grade III lesions in the study sample. They were intubated and hemodynamically stabilized at admission to minimize any confounding biases resulting from hypoxia and hypotension regarding the neurological outcomes. CT of the head was performed to rule out any possible posttraumatic mass lesions. Patients with significant polytrauma and thereby hemodynamically unstable were excluded from the study.

T2-weighted images, obtained in 3 Tesla MRI, were independently assessed by the authors for the evidence of hypointense signals in gradient recall echo and weighted imaging sequences or restricted diffusion in DWI sequences. Only the patients with significant lesions in the brainstem were included. Patients with any additional lesions either in the corpus callosum or in the anterior or posterior commissures were excluded. We stratified the brainstem into (1) parts: midbrain, pons, medulla and (2) locations: anterior part encompassing the corticospinal tracts and posterior part mainly dealing with maintaining arousal via the reticular activating system and the autonomic control system (damage predisposing to dysautonomia) (Figs. 1 and 2). We determined the differences between the lesion and the corresponding half in fractional anisotropy. We also calculated the differences between the DWI value in the rostral-caudal region along the corticospinal tracts and the affected tracts and the values obtained from the seeds placed in the region of interest (ROI). The acquisition of DTI was done to visualize any compression, discontinuity, abrupt termination, or tract wasting. The number of days it took from brain damage to spontaneous and sustained eye opening and the time required for weight-bearing on a dysfunctional body part, with minimal external support, were recorded for each patient. Stringent care for recognizing early features of dysautonomia was also carried out in all patients and was managed according to the treatment protocol. The number of days to early recovery was formulated in terms of fractional anisotropy and DWI values, taking into account also the location of a lesion.

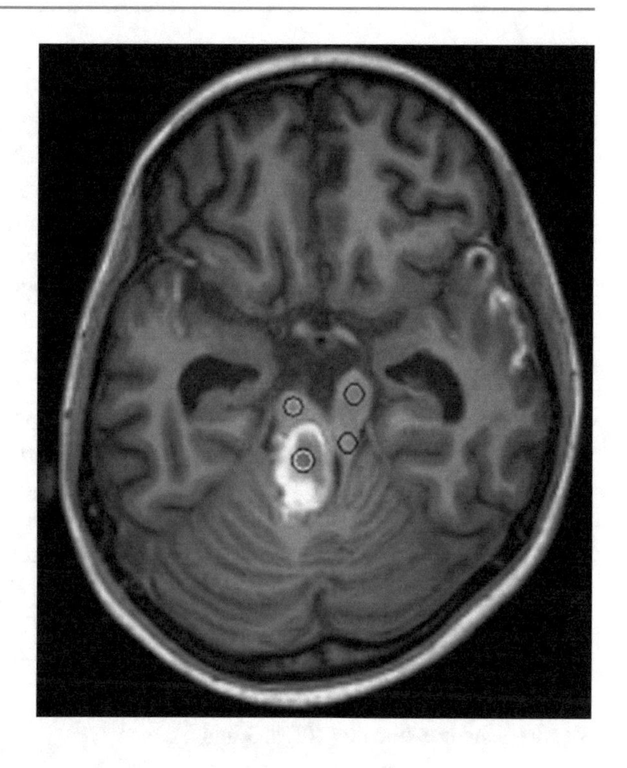

Fig. 1 Axial fluid-attenuated inverse recovery (FLAIR) MRI sequence at the midbrain level wherein the corticospinal tracts are depicted in the anterior (upper circles) and posterior (lower circles) locations of the brainstem

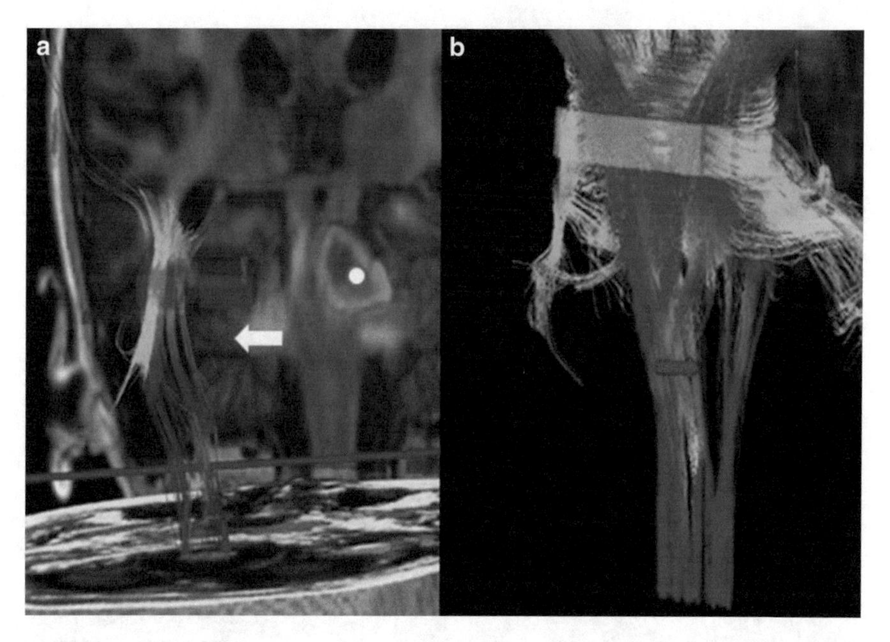

Fig. 2 (**a**) MRI tractography depicting the absence of tract fibers (white arrow) below the level of a pontile lesion (white circle); (**b**) a blown-up image of the tract fibers shown in Panel "a" – red arrow depicts wasting of the tract fibers compared to the normal counterpart

The results were analyzed by creating graphs correlating power trend line in fractional anisotropy and by obtaining the DWI values in determining the number of days to recovery. Further, correlation graphs through bivariate analysis and curve estimation by regression analysis were created for depicting the number of days to recovery in case of lesions with anterior or posterior brain stem localization and also in patients with dysautonomia. The analysis was performed with the help of SPSS v16 (SPSS Inc., Chicago, IL) for conducting regression analysis.

3 Results

The power trend lines pointed to a positive correlation between the increasing DWI values along the corticospinal tracts and increasing number of days for clinical improvement with regards to standing and eye-opening (Figs. 3 and 4).

Bivariate correlation analysis showed that only DWI significantly correlated with the number of days required for early standing and eye-opening (Table 1).

A curve estimation by regression analysis showed that only DWI values positively correlated with the prediction of a number of days for early recovery in terms of eye-opening and standing with minimal support among our patients with DAI grade III injury (Figs. 5 and 6). The estimation graphs also showed that the prediction of onset of dysautonomic features correlated with the anterior (corticospinal tract region) and posterior (axial portion posterior to the corticospinal tract region) subdivisions of the brainstem (Fig. 7).

The average number of days until regaining the ability for standing with support in patients with anterior corticospinal lesions was 10.4 days, whereas it was 13.3 days for those with posterior-based lesions. In patients with dysautonomic features, it took, on average, 32.6 days for early

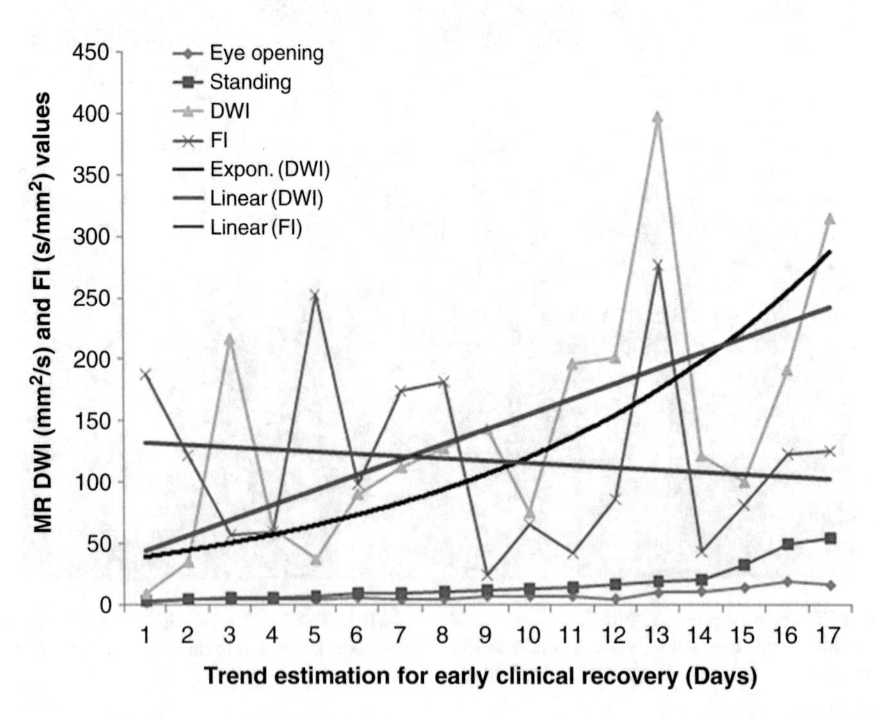

Fig. 3 Estimations of linear and exponential trend lines for diffusion-weighted (DWI) and fractional anisotropy (FI) magnetic resonance imaging in terms of days required for sustained eye-opening and standing with minimal support in patients with diffuse axonal injury (DAI) grade III

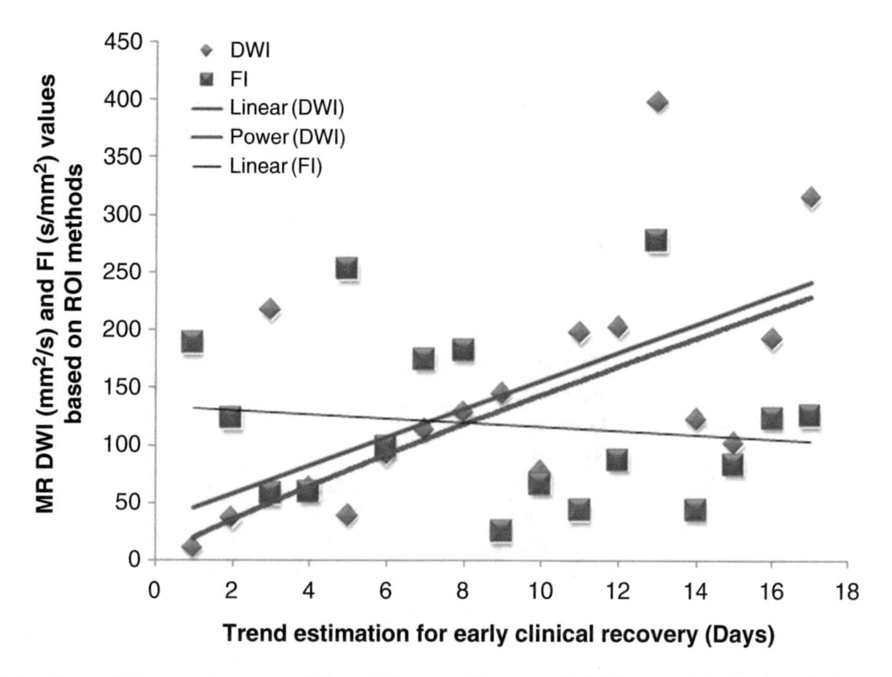

Fig. 4 Estimations of linear and exponential trend lines for diffusion-weighted (DWI) and fractional anisotropy (FI) magnetic resonance imaging concerning the regions of interest (ROI) in terms of days required for sustained eye-opening and standing with minimal support in patients with diffuse axonal injury (DAI) grade III

clinical recovery, which was significantly longer than in the two aforementioned lesions ($p < 0.05$).

4 Discussion

Diffuse axonal injury (DAI) is the most common cause of posttraumatic coma, disability, and persistent vegetative state (Gennarelli 1987). It is a consequence of stretch, twist, and tearing of axons and their microvasculature, or of shearing forces, resulting from accelerating and deaccelerating mechanisms. DAI is defined as a state of prolonged coma beyond 6 h of injury despite any obvious demonstrable lesions of the intracranial mass (Eum et al. 1998; Alberico et al. 1987). Such injury has been first described by Strich (1956) as a widespread degeneration of the white matter in a specific subset of patients with posttraumatic dementia. The time course of pathological changes has been later depicted by Adams et al. (1977). Histological markers include the areas of focal hemorrhage and the emergence of axonal retraction balls (Ma et al. 2016). Subsequently, microglial clusters appear and the axonal retraction balls start to disappear. Months after the injury, there is a wasting of the white matter tract owing to degeneration (Gennarelli et al. 1982).

Computed tomography lacks accuracy in predicting outcome and does not correlate well with the Glasgow Coma Scale (GCS) score or neurological status (Kim et al. 2001; Zimmerman et al. 1986). Based on MRI imaging, patients with pure DAI are divided into three grades according to the Adams anatomical grading system (Adams et al. 1989). Grade I is defined as lesions seen in the white matter of a cerebral hemisphere, grade II as lesions in the corpus callosum, and grade III as lesions in the brain stem (Park et al. 2009; Mac Donald et al. 2007). In this study, we divided the brainstem into four parts in the rostro-caudal direction as midbrain, pons, medulla, and lastly as combined. Further, we divided each part into the anterior location containing the corticospinal tracts and the rest as the posterior location in the axial sections.

Table 1 Bivariate Kendall's tau_b and Spearman's rho correlation analyses of diffusion-weighted image (DWI) and fractional anisotropy (FI) of magnetic resonance for predicting early clinical recovery marked by eye-opening and standing with minimal support in patients with diffuse axonal injury (DAI) grade III

			Eye-opening	Standing	DWI	FI
Kendall's tau_b	Eye-opening	Correlation coefficient	1	0.781	0.339	0.016
		Significance		0.001	0.066	0.933
	Standing	Correlation coefficient	0.781	1	0.509	0.052
		Significance	0.001		0.004	0.772
	DWI	Correlation coefficient	0.339	0.509		1
		Significance	0.066	0.004		1
	FI	Correlation coefficient	0.016	0.052	0.001	1
		Significance	0.933	0.772	1	
Spearman's rho	Eye-opening	Correlation coefficient	1	0.890	0.451	0.016
		Significance		0.001	0.069	0.953
	Standing	Correlation coefficient	0.890	1	0.616	0.020
		Significance	0.001		0.009	0.940
	DWI	Correlation coefficient	0.451	0.616	1	0.021
		Significance	0.069	0.009		0.094
	FI	Correlation coefficient	0.016	0.020	0.021	1
		Significance	0.953	0.940	0.937	

Significance of correlation is at $p < 0.01$ level (two-tailed)

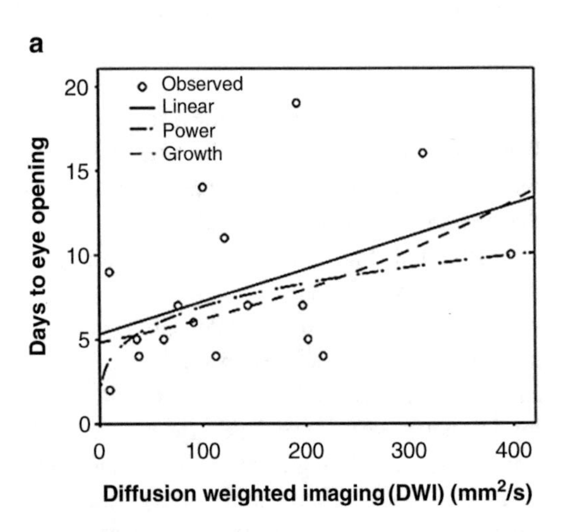

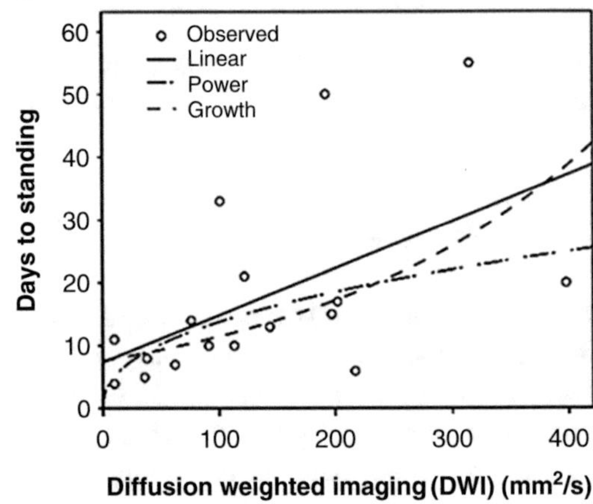

Fig. 5 The observed, linear, power, growth, and exponential curve estimations of diffusion-weighted (DWI) magnetic resonance imaging for prediction of days required for early recovery marked by eye-opening (**a**) and standing with minimal support (**b**) in patients with diffuse axonal injury (DAI) grade III

A previous study has shown a trend for an increasing number of days required for recovery with higher grades of DAI wherein patients with DAI grade III take, on average, almost 2 months for regaining awake state (Park et al. 2009). A correlation has been shown between higher DAI grades, noticed in MRI, and a longer duration of loss of consciousness, as well as between higher grades of lesions and worse outcome (Calvi et al. 2011; Skandsen et al. 2010; Fabbri et al. 2008; Kim et al. 1997). In the present study, the average number of days required for standing with support was 10.4 for the anterior lesions, 13.3 for the posterior lesions, and 32.6 for patients with dysautonomic features. Oh et al. (2001) have demonstrated a similar trend in outcome in higher

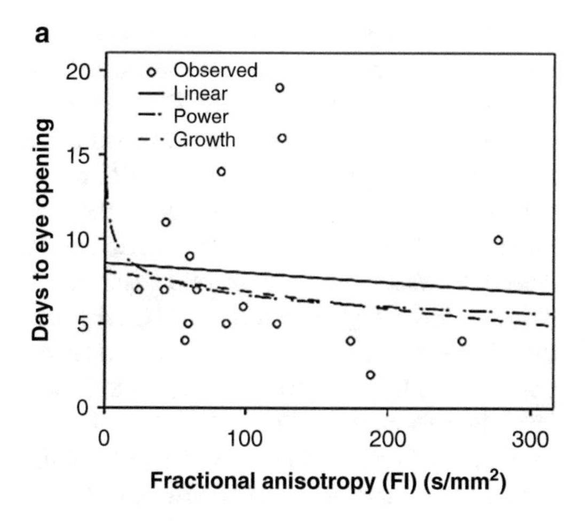

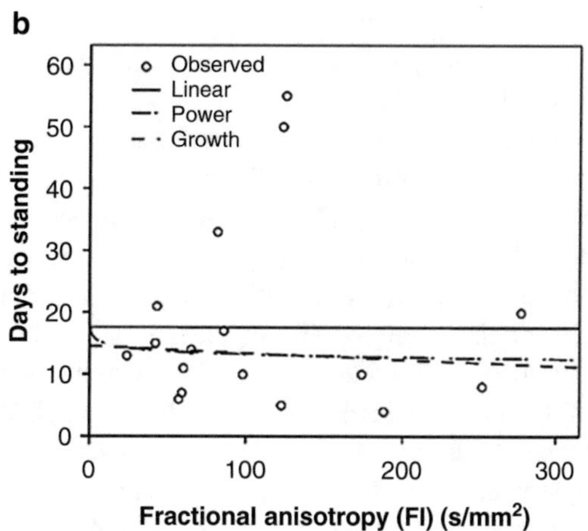

Fig. 6 The observed, linear, power, growth, and exponential curve estimations of fractional anisotropy (FI) magnetic resonance imaging for prediction of days required for early recovery marked by eye-opening (**a**) and standing with minimal support (**b**) in patients with diffuse axonal injury (DAI) grade III

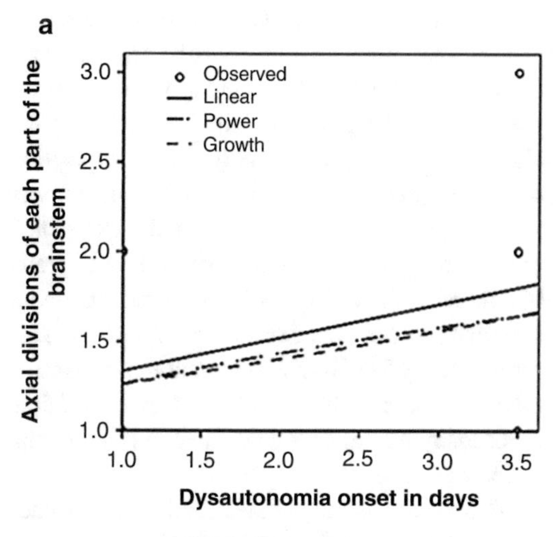

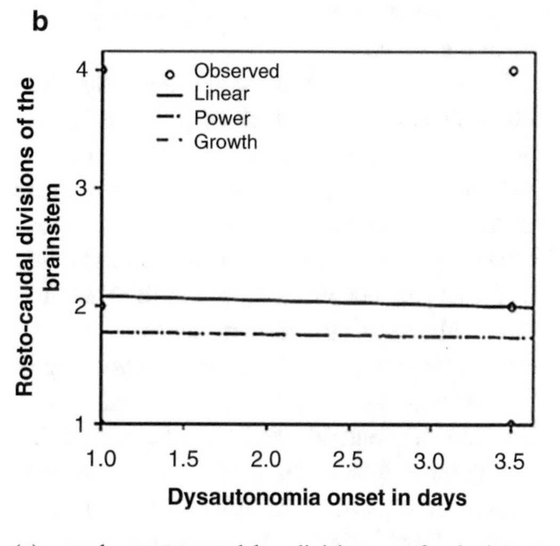

Fig. 7 The observed, linear, power, growth, and exponential curve estimation for prediction of days to the onset of dysautonomia in axial divisions (anterior vs. posterior) (**a**) and rostro-caudal divisions of brainstem (midbrain vs. pons vs. medulla) (**b**) in patients with diffuse axonal injury (DAI) grade III

grade lesions, however with no correlation between the lesion sizes and outcome. Park et al. (1991) have found that almost one half of the patients with DAI grades II–III do not regain consciousness compared to around 15% of patients with grade 1 lesion. In the present study, most of the patients eventually had early good recovery in terms of spontaneous and sustained eye-opening and minimal weight support bearing; except two patients with severe dysautonomia, one of whom was eventually taken home against medical advice and another deceasing after a prolonged hospital stay. Some studies have reported that the severity of DAI, the length of hospital stay, or the absence of consciousness recovery are predictive of both mortality and dependency (Vieira et al. 2016; Chelly et al. 2011). Further, multivariate analysis has revealed independent relationships between poor outcome and lesions in the mesencephalic region corresponding to substantia nigra and tegmentum present in weighted imaging and with age (Abu Hamdeh et al. 2017).

Autonomic dysregulation is generally an independent prognostic factor (Brezova et al. 2014; Sidaros et al. 2008). Dysautonomia is associated with diffuse axonal injury (DAI) and the development of spasticity, resulting in poor outcome (Hendricks et al. 2010; Baguley et al. 1999). Posteriorly located lesions correlate with the risk of dysautonomia and thereby poorer neurological outcome. In the present study, the appearance of dysautonomia was the only significant predictor of mortality. We noticed two cases of DAI grade III with the involvement of the thalamus; both were complicated with dysautonomia. Our patients with dysautonomia had the greatest increase in the number of days needed for recovery (average of 32.6 days for weight bearing) and one of them died.

The unfavorable outcome increases threefold with each MRI grade, which helps make clinical decision and provide information to patients with DAI and to their families. A higher DAI grade results in a greater risk of unfavorable outcome. Noticeably, however, favorable outcome is found in 37% of patients with DAI grade III. The number or volume of DAI lesions seems not predictive of outcome. In a cohort study comprising of 177 DAI patients, more than one half had a favorable functional outcome and a good quality of life at long-term follow-up, which also concerned some patients with DAI grade III (van Eijck et al. 2018a). Nonetheless, Firsching et al. (2002) have reported mortality in patients with DAI in relation to MRI grading and found that only 1 out of the 32 patients with DAI grade I–II died as opposed to 8 with DAI grade III, resulting in OR of 10.3 (Firsching et al. 2002). In the present study, 88.2% of patients eventually made a good early recovery, despite being diagnosed as DAI grade III, and the mortality rate was just 5.9%.

Diffusion tensor imaging (DTI), which measures the degree of anisotropy of water diffusion in white matter fibers, provides information on the nerve alignment and integrity (Ma et al. 2016; Inglese et al. 2005). Mac Donald et al. (2007) have suggested that diffusion tensor imaging (DTI) is more sensitive to evaluate DAI lesions than conventional MRI. Gradient recall echo and diffusion-weighted imaging can detect minor hemorrhages; thereby, such MRI modes are suitable for the diagnosis of DAI in the early phase. Ashwal et al. (2006) have reported that the detection rate of hemorrhagic lesions in susceptibility-weighted imaging is sixfold greater than that in conventional MRI. Acute DAI lesions unraveled in DTI may be stable for up to 5 years postinjury as indicated by some studies (Dinkel et al. 2014). Other studies, however, consent that both hemorrhagic and nonhemorrhagic lesions wane over time, so that prognostic information may be lost during late MRI use (Moen et al. 2012; Messori et al. 2003). Since early MRI can be challenging to perform in severely traumatized patients with brain injury, there is a paucity of studies that would evaluate MRI in the early phase of DAI-induced traumatic brain injury for prediction of long-term outcome, which was the chief goal of the present study. Early acquisition of MRI sequences would help predict the number of days to possible recovery and the associated risk of complications in patients with DAI grade III, thereby providing fuller information about outcome. However, a decision of whether to perform an early MRI ought to be cautiously taken as DAI patients often suffer from multiple accompanying injuries and have unstable vital signs (Lee et al. 2018).

This study demonstrates that DWI values along the affected corticospinal tracts were related to the increased number of days required for early clinical recovery in patients with DAI grade III. Furthermore, posterior location of lesions was associated with dysautonomia, thereby extending the time to recovery and increasing mortality. We conclude that MRI tractography has a potential of becoming a surrogate marker for effectively predicting days to recovery in diffuse axonal injury grade III, which may have an essential value in setting an effective therapeutic algorithm for managing such patients.

Conflicts of Interest The authors declare no conflicts of interest in relation to this article.

Ethical Approval All procedures performed in studies involving human participants were in accordance with the ethical standards of the institutional and/or national research committee and with the 1964 Helsinki declaration and its later amendments or comparable ethical standards. The study was approved by the Ethics Committee of the Nobel Medical College and Teaching Hospital in Biratnagar, Nepal (IRC-NMCTH-169/2018).

Informed Consent Written informed consent was obtained from the guardians, relatives, or next of kin concerning all individual participants included in the study.

References

Abu Hamdeh S, Marklund N, Lannsjö M, Howells T, Raininko R, Wikström J, Enblad P (2017) Extended anatomical grading in diffuse axonal injury using MRI: hemorrhagic lesions in the substantia nigra and mesencephalic tegmentum indicate poor long-term outcome. J Neurotrauma 34(2):341–352

Adams H, Mitchell DE, Graham DI, Doyle D (1977) Diffuse brain damage of immediate impact type. Its relationship to "primary brain stem damage" in head injury. Brain 100:489–502

Adams JH, Doyle D, Ford I, Gennarelli TA, Graham DI, McLellan DR (1989) Diffuse axonal injury in head injury: definition, diagnosis and grading. Histopathology 15:49–59

Alberico AM, Ward JD, Choi SC, Marmarou A, Young HF (1987) Outcome after severe head injury: relationship to mass lesions, diffuse injury, and ICP course in pediatric and adult patients. J Neurosurg 67:648–656

Ashwal S, Babikian T, Gardner-Nichols J, Freier MC, Tong KA, Holshouser BA (2006) Susceptibility-weighted imaging and proton magnetic resonance spectroscopy in assessment of outcome after pediatric traumatic brain injury. Arch Phys Med Rehabil 87 (12 Suppl 2):S50–S58

Baguley IJ, Nicholls JL, Felmingham KL, Crooks J, Gurka JA, Wade LD (1999) Dysautonomia after traumatic brain injury: a forgotten syndrome? J Neurol Neurosurg Psychiatry 67(1):39–43

Brezova V, Moen KG, Skandsen T, Vik A, Brewer JB, Salvesen O, Haberg AK (2014) Prospective longitudinal MRI study of brain volumes and diffusion changes during the first year after moderate to severe traumatic brain injury. Neuroimage Clin 5:128–140

Calvi MR, Beretta L, Dell'Acqua A, Anzalone N, Licini G, Gemma M (2011) Early prognosis after severe traumatic brain injury with minor or absent computed tomography scan lesions. J Trauma 70 (2):447–451

Chelly H, Chaari A, Daoud E, Dammak H, Medhioub F, Mnif J et al (2011) Diffuse axonal injury in patients with head injuries: an epidemiologic and prognosis study of 124 cases. J Trauma 71(4):838–846

Dinkel J, Drier A, Khalilzadeh O, Perlbarg V, Czernecki V, Gupta R, Gomas F, Sanchez P, Dormont D, Galanaud D, Stevens RD, Puybasset L, for NICER (Neuro Imaging for Coma Emergence and Recovery) Consortium (2014) Long-term white matter changes after severe traumatic brain injury: a 5-year prospective cohort. Am J Neuroradiol 35:23–29

Eum SW, Lim DJ, Kim BR, Cho TH, Park JY, Suh JK et al (1998) Prognostic factors in patients with diffuse axonal injury. J Korean Neurosurg Soc 27:1668–1674

Fabbri A, Servadei F, Marchesini G, Stein SC, Vandelli A (2008) Early predictors of unfavorable outcome in subjects with moderate head injury in the emergency department. J Neurol Neurosurg Psychiatry 79 (5):567–573

Firsching R, Woischneck D, Klein S, Ludwig K, Dohring W (2002) Brain stem lesions after head injury. Neurol Res 24(2):145–146

Gennarelli TA (1987) Cerebral concussion and diffuse brain injuries. In: Cooper PR (ed) Head injury, 2nd edn. Williams & Wilkins, Baltimore, pp 108–124

Gennarelli T, Thibault L, Adams J, Graham D, Thompson C, Marcincin R (1982) Diffuse axonal injury and traumatic coma in the primate. Ann Neurol 12:564–574

Hendricks HT, Heeren AH, Vos PE (2010) Dysautonomia after severe traumatic brain injury. Eur J Neurol 17 (9):1172–1177

Inglese M, Makani S, Johnson G, Cohen BA, Silver JA, Gonen O, Grossman RI (2005) Diffuse axonal injury in mild traumatic brain injury: a diffusion tensor imaging study. J Neurosurg 103:298–303

Kim CH, Lee HK, Koh YC, Hwang DY (1997) Clinical analysis of diffuse axonal injury (DAI) diagnosed with magnetic resonance image (MRI). J Korean Neurosurg Soc 26:241–248

Kim HJ, Park IS, Kim JH, Kim KJ, Hwang SH, Kim ES et al (2001) Clinical analysis of the prognosis of the patients with cerebral diffuse axonal injuries, based on gradient-echo MR imaging. J Korean Neurosurg Soc 30:168–172

Lee HJ, Sun HW, Lee SJ, Choi NJ, Jung YJ, Hong SK (2018) Clinical outcomes of diffuse axonal injury according to radiological grade. J Trauma Inj 31 (2):51–57

Ma J, Zhang K, Wang Z, Chen G (2016) Progress of research on diffuse axonal injury after traumatic brain injury. Neural Plast 2016:9746313

Mac Donald CL, Dikranian K, Song SK, Bayly PV, Holtzman DM, Brody DL (2007) Detection of traumatic axonal injury with diffusion tensor imaging in a mouse model of traumatic brain injury. Exp Neurol 205:116–131

Messori A, Polonara G, Mabiglia C, Salvolini U (2003) Is haemosiderin visible indefinitely on gradient-echo MRI following traumatic intracerebral haemorrhage? Neuroradiology 45:881–886

Moen KG, Skandsen T, Folvik M, Brezova V, Kvistad KA, Rydland J, Manley GT, Vik A (2012) A longitudinal MRI study of traumatic axonal injury in patients with moderate and severe traumatic brain injury. J Neurol Neurosurg Psychiatry 83(12):1193–1200

Oh KS, Ha SI, Suh BS, Lee HS, Lee JS (2001) The correlation of MRI findings to outcome in diffuse axonal injury patients. J Korean Neurosurg Soc 30 (Suppl I):S20–S24

Park SW, Park K, Kim YB, Min BK, Hwang SN, Suk JS et al (1991) Prognostic factors in diffuse axonal injuries of brain. J Korean Neurosurg Soc 20:983–990

Park SJ, Hur JW, Kwon KY, Rhee JJ, Lee JW, Lee HK (2009) Time to recover consciousness in patients with diffuse axonal injury: assessment with reference to magnetic resonance grading. J Korean Neurosurg Soc 46(3):205–209

Sidaros A, Engberg AW, Sidaros K, Liptrot MG, Herning M, Petersen P, Paulson OB, Jernigan TL, Rostrup E (2008) Diffusion tensor imaging during recovery from severe traumatic brain injury and relation to clinical outcome: a longitudinal study. Brain 131:559–572

Skandsen T, Kvistad KA, Solheim O, Strand IH, Folvik M, Vik A (2010) Prevalence and impact of diffuse axonal injury in patients with moderate and severe head injury: a cohort study of early magnetic resonance imaging findings and 1-year outcome. J Neurosurg 113(3):556–563

Strich SJ (1956) Diffuse degeneration of the cerebral white matter in severe dementia following head injury. J Neurol Neurosurg Psychiatry 19:163–185

van Eijck MM, Schoonman GG, van der Naalt J, de Vries J, Roks G (2018a) Diffuse axonal injury after traumatic brain injury is a prognostic factor for functional outcome: a systematic review and meta-analysis. Brain Inj 32(4):395–402

van Eijck M, van der Naalt J, de Jongh M, Schoonman G, Oldenbeuving A, Peluso J, de Vries J, Roks G (2018b) Patients with diffuse axonal injury can recover to a favorable long-term functional and quality of life outcome. J Neurotrauma 35(20):2357–2364

Vieira RC, Paiva WS, de Oliveira DV, Teixeira MJ, de Andrade AF, de Sousa RM (2016) Diffuse axonal injury: epidemiology, outcome and associated risk factors. Front Neurol 7:178

Zimmerman RA, Bilaniuk LT, Hackney DB, Goldberg HI, Grossman RI (1986) Head injury: early results of comparing CT and high-field MR. Am J Roentgenol 147:1215–1222

Adv Exp Med Biol - Clinical and Experimental Biomedicine (2020) 8: 29–38
https://doi.org/10.1007/5584_2019_465

Published online: 14 January 2020

Integrated Thermal Rehabilitation: A New Therapeutic Approach for Disabilities

Giovanni Barassi, Esteban Obrero-Gaitan, Giuseppe Irace, Matteo Crudeli, Giovanni Campobasso, Francesco Palano, Leonardo Trivisano, and Vito Piazzolla

Abstract

A mutual link between somatic and visceral neural pathways is known in medicine. This study addresses therapeutic effectiveness of an integrated rehabilitative approach of thermal aquatic environment, combined with neuromuscular manual stimulation, on activation of afferent sensory visceral and somatic efferent neuronal motor pathways in different pathologies of neuromuscular motor and respiratory systems. The study included 63 patients subjected to a protocol consisting of hydroponic treatment, hydrokinesitherapy associated with ozonized vascular pathway, and mud therapy associated with cardiorespiratory treatment performed in aquatic environment and aided by neuromuscular manual therapy. The therapeutic protocol consisted of rehabilitation sessions 5 days a week for 2 months. The outcome measures were spirometry tests and the following evaluation instruments: Tinetti Gait and Balance assessment scale, Functional Independence Measurement, visual analogue scale, and the EQ-5D-5 L instrument. The tests were applied before and after the protocol completion. The findings demonstrate a general increase in patients' everyday living autonomy and quality of life, with a particular improvement in respiratory function tests. We conclude that the integrated thermal approach holds promise in therapeutic rehabilitation of disabilities.

Keywords

Balneotherapy · Disability · Hydrokinesitherapy · Neuromuscular manual therapy · Physiotherapy · Respiratory system · Spirometry · Thermal therapy

1 Introduction

Thermal rehabilitative therapy, a branch of balneotherapy, is a valid therapeutic choice in rheumatological, musculoskeletal, and

G. Barassi (✉)
Department of Medical and Oral Sciences and Biotechnologies, "Gabriele d' Annunzio" University Chieti-Pescara, Chieti, Italy
e-mail: coordftgb@unich.it; dottgiovannibarassi@gmail.com

E. Obrero-Gaitan
Faculty of Health Sciences, Jaen University, Jaen, Spain

G. Irace, M. Crudeli, and F. Palano
Thermal Medicine Center of Castelnuovo della Daunia, Foggia, Italy

G. Campobasso
Department of Health, Social Wellness and Sport Promotion, Puglia Region, Bari, Italy

L. Trivisano
Rehabilitation Department, Local Health Unit of Foggia, Foggia, Italy

V. Piazzolla
Local Health Unit of Foggia, Foggia, Italy

neurological diseases. This branch of rehabilitation medicine gained popularity in the eighteenth and nineteenth centuries due its diuretic effect paired with the excretion of sodium, potassium, calcium, and lead (Pappas and Perlman 2002). Contemporary studies have extended the positive effects of thermal therapy to rheumatologic and neurologic disorders. That may be exemplified by osteoarthritis or cerebral palsy, where this therapy is particularly useful for spasticity treatment and motion improvement, creating an advantageous environment for adapted physical activity in psychomotor retardation (Fioravanti et al. 2012; Pospíšil et al. 2007; (Verhagen et al. 2007; Holdcraft et al. 2003; Hutzler et al. 1998; Jochheim 1990). Any rehabilitative act should consider the complexity of the human body system, trying to identify the possible connections between different dysfunctions, which requires individually tailored rehabilitative protocols. Thermal aquatic therapy has an array of therapeutic indications and thus may be integrated into different rehabilitative techniques that are offered to patients.

Medical literature describes mutuality of the somatic and visceral pathways. In particular, it has been demonstrated how neuromuscular manual treatment can replace pharmacological treatment or be integrated with it, improving beneficial effects. Barassi et al. (2018b) have shown that neuromuscular manual stimulation improves motor control. Such improvement plays a role in prevention of dysfunction caused by chronic pain, since a mismatch between motor activity and sensory feedback increases the activity of nociceptors in the central nervous system.

Hydrokinesitherapy is also effectively applied to patients with painful neurological or musculoskeletal alterations due to the warmth and floatability of water, able to inhibit nociceptors by acting on thermal receptors and mechanoreceptors. Water attributes enable the kind of exercise, such as articular techniques or gait and balance training, which would not be possible in the terrestrial environment. Thermal therapy, as alternative musculoskeletal and rheumatologic treatment, may assume the form of an integrated protocol composed by local mudpack therapy, hydrotherapy with mineral water, and humid hot inhalations (Staffieri and Abramo 2007). A myriad of experiments performed in the last two centuries have attested to the effectiveness of spa treatment in enhancing health, mood, and quality of life (Walsh and Gerley 1985). This treatment, despite not being universally accepted as part of evidence-based medicine, ought to be performed in a controlled and supervised manner on par with other clinical treatments (Yurtkuran et al. 1999). A complementary branch of thermal therapy, particularly useful for visceral stimulation, is thermal water drinking with its capability to strengthen the innate antioxidant system, whose dysfunction underlies chronic inflammatory disorders (Constant et al. 1995; Konrad et al. 1992; Langlais et al. 1991).

According to the philosophy of bioprogressive rehabilitation, combining different types of physiotherapy may potentiate beneficial outcomes, leading to a reduction in clinical symptoms (Pokorski et al. 2018; Saggini et al. 2017; Hurley et al. 2012). There are just few studies on the integrated rehabilitative approach consisting of thermal therapy and specific physiotherapy (Dogaru et al. 2017; Bellomo et al. 2015; Bellomo et al. 2016). Therefore, in this study, we investigated the effects of therapy, named integrated thermal care (ITC), consisting of activation of both visceral and somatic pathways, in neuromuscular pathologies.

2 Methods

2.1 Patients and Study Protocol

This study was designed in a local healthcare company (ASL FG, Italy) and was conducted in the Thermal Medicine Conventioned Center in the town of Castelnuovo della Daunia in Italy. There were 63 patients (F/M; 29/34) of the mean age 55 ± 21 years recruited for the study, who suffered from respiratory and neuromotor pathologies. All of the patients underwent a general medical and cognitive examination. We

Table 1 Sociodemographic characteristics of patients

	Sample	Age		Gender	
				Female	Male
Pathological condition	n	Mean ± SD	Min–max	n (%)	n (%)
CP	17	54 ± 22	11–84	6 (35)	11 (65%)
OA	13	61 ± 17	27–82	7 (54)	6 (46)
RD	16	62 ± 17	23–92	9 (56)	7 (44)
ND	17	45 ± 23	10–81	7 (41)	10 (59)
Total	63	55 ± 21	10–92	29 (46)	34 (54)

CP cerebral palsy, *OA* osteoarthritis, *RD* respiratory disease, *ND* neurological motor and cognitive disorders

stratified the patients by the type of pathology into the following four groups (Table 1):

- Cerebral palsy – 17 patients
- Osteoarthritis – 13 patients
- Respiratory disorders – 16 patients
- Neurological motor and cognitive disorders of neurodegenerative origin – 17 patients

Inclusion criteria were as follows:

- Neuromotor disability of 67%, the threshold disability level in the healthcare system in Italy
- Associated pain symptoms
- Attendance at the thermal facilities where the research was conducted.

Exclusion criteria were as follows:

- Uncompensated heart disease
- Acute musculoskeletal diseases
- Uncontrollable fecal or urinary incontinence
- Multiple sclerosis
- Current infections and mycosis
- Fever.

The therapeutic protocol was composed of hydroponic treatment daily, hydrokinesitherapy associated with ozonized vascular pathway 3 days a week, bathing in alkaline bicarbonate groundwater, and mud therapy associated with cardiorespiratory treatment. Rehabilitative sessions were conducted 5 days a week for 2 months. Neuromuscular manual therapy (NMT) was conducted twice a week. Hydroponic treatment consisted of individual intake of 0.5–1.2 L of mineral water coming from local underground springs. The bicarbonate-sulfate-alkaline-bromine-iodide-containing waters have spasmolytic properties often useful in the treatment of gastrointestinal ailments. Hydrokinesitherapy consisted of peripheral sensorineural stimulation associated with NMT in the aquatic environment, used for motor re-education (Barassi et al. 2019). Each hydrokinesitherapy session included the following:

- 5 min of adaptation and relaxation
- 10 min of walking
- 10 min of joint mobility exercise for improvement of proprioception and to increase muscle strength
- 15 min of stretching.

The ozonized vascular pathway was performed in tanks filled with water at temp. 24 °C and 38 °C enriched with ozone, equipped with pressure water jets of 4–6 atm (Restaino et al. 1995). This treatment exerts the alternating contracting-relaxing influence on blood vessels during a 20-min water walk (Burton and Taylor 1940). Moreover, ozone present in the water has an antimicrobial effect on the skin, which helps maintain personal care and hygiene (Travagli et al. 2010).

NMT was applied to the body regions known for the involvement in the control of posture, pain, and the autonomic nervous system (Barassi et al. 2018a). The goal of this treatment was to rebalance the afferent discharge transmitted into the brain, to improve the motility of muscular structures, and in effect to help maintain the body homeostasis. The choice of muscles to be treated was related to the concept of specific muscle dysfunction (Barassi et al. 2019; Bellomo et al. 2015; Travell and Simons 1983). The principal muscles treated were as follows:

- Iliopsoas
- Quadratus plantae
- Diaphragm
- Quadratus lumborum
- Pectoralis minor
- Levator scapulae.

In addition, patients were subjected to a 20-min-long application of matured mud in association with NMT. Mud treatment was used due to its capability of influencing chondrocyte activity. This treatment may decrease the activity of proinflammatory cytokines, such as interleukin 1 and tumor necrosis factor, which play a key role in osteoarthritic joint degeneration (Grunke and Schulze-Koops 2006; Bellometti et al. 1997). Patients also had a 60-min-long cycle of hot-humid inhalation in each rehabilitative session to mitigate the influence of the obstructive component of lung function (Stein and Thiel 2017).

2.2 Outcome Measures

The outcome measures investigated at baseline (pre) and after 2-month-long rehabilitative treatment (post) were as follows:

- Spirometry variables to evaluate lung function: forced vital capacity (FVC), forced expiration volume in 1 s (FEV1), and peak expiratory flow (PEF). The best result of three trials of each variable was taken into further data evaluation and was expressed as percent of predictive value.
- Tinetti Gait and Balance (TIN) scale to evaluate the patient's balance stability and motor control (Sauvage et al. 1992)
- Functional Independence Measure (FIM) questionnaire to evaluate the patient's daily life and health-related status, in particular motility, sphincter control, motor control, and social and cognitive communication (Ravaud et al. 1999)
- Visual analogue scale (VAS) to evaluate pain perception, with 0 denoting no pain and 10 denoting maximum pain
- EuroQol Scale (EQ-5D-5 L) to evaluate health-related quality of life (Herdman et al. 2011).

2.3 Statistical Elaboration

Data are continuous and expressed as means ±SD. Data distribution was checked with the Shapiro-Wilk test. Significance of differences between the means of related samples pre- vs. posttreatment was checked with a *t*-test for normally distributed data or the Wilcoxon test for abnormal distribution. A p-values <0.05 defined statistically significant differences. Variations between pre- and post-measurement values were estimated. The effect size of the intervention was estimated as a standardized mean difference (d-Cohen). Values $d = 0.2$ and below were considered a small effect size, $d = 0.5$ a medium effect size, and 0.8 and more a large effect size. That is to mean that if the means of two groups differed by less than 0.2SD, the difference was considered of no real importance, even if statistically significant. A commercial statistical package SPSS v21.0 for Windows was used in the analysis (IBM Corp, Armonk, NY).

3 Results

3.1 Cerebral Palsy

Patients with cerebral palsy experienced improvements in all of the outcome measures after the 5-month-long rehabilitative intervention

Table 2 Outcome measures of rehabilitative interventions in cerebral palsy patients

	Intervention – mean (SD)				Statistics			Effect size
Outcome	Pre	Post	Δ Difference	Variation	Distribution	Test	p-value	d-Cohen
FVC (L)	2.6 (1.3)	2.9 (1.8)	0.3 (0.8)	11.1%	Abnormal	Wilcoxon	0.118	0.19
FVC (% predicted)	76.3 (22.8)	83.2 (27.8)	7.0 (16.9)	9.2%	Abnormal	Wilcoxon	0.079	0.27
FEV1 (L)	2.4 (1.2)	2.5 (1.0)	0.1 (0.3)	4.7%	Abnormal	Wilcoxon	0.079	0.09
FEV1 (% predicted)	83.8 (25.2)	90.2 (22.8)	6.4 (12.5)	7.6%	Abnormal	Wilcoxon	0.055	0.27
PEF (L/min)	4.2 (2.1)	5.2 (1.8)	1.0 (0.8)	22.7%	Normal	*t*-test	<0.001	0.48
PEF (% predicted)	57.8 (20.6)	74.1 (18.0)	16.4 (14.8)	28.3%	Normal	*t*-test	<0.001	0.84
FIM (score)	90.2 (18.9)	95.0 (22.0)	4.8 (16.4)	5.4%	Normal	*t*-test	0.242	0.23
TIN (score)	10.5 (7.9)	12.5 (8.9)	2.0 (3.5)	19.2%	Normal	*t*-test	0.030	0.24
VAS (score)	5.9 (2.8)	4.1 (2.3)	−1.8 (1.4)	31.0%	Normal	*t*-test	<0.001	0.71
EQ-5D-5 L (score)	53.0 (21.6)	64.4 (19.4)	11.4 (11.8)	21.5%	Normal	*t*-test	0.001	0.55

FVC forced vital capacity, *FEV1* forced expiratory volume in 1 s, *PEF* peak expiratory flow, *FIM* Functional Independence Measure scale, *TIN* Tinetti Gait and Balance scale, *VAS* visual analogue scale for pain perception, *EQ-5D-5 L* EuroQol health scale

(Table 2). The improvements reached statistical significance in spirometry PEF absolute values ($p < 0.001$; effect size $d = 0.48$) and %PEF predicted values ($p < 0.001$; $d = 0.84$), Tinetti scale ($p = 0.030$; $d = 0.24$), and general health ($p < 0.001$; $d = 0.55$), with concomitant decrease in pain perception ($p < 0.001$; $d = 0.71$).

3.2 Osteoarthritis

In the patients with osteoarthritis, a trend toward improvement in clinical condition was clearly noticeable (Table 3), except for lung function variables that remained unchanged after rehabilitation. However, the scales describing daily functioning, quality of life, and neuromuscular stability significantly improved posttreatment: FIM ($p = 0.012$; effect size $d = 0.47$), Tinetti scale ($p = 0.003$; $d = 0.88$), and general health perception ($p < 0.001$). Pain perception decreased ($p < 0.001$; $d = 1.07$).

3.3 Respiratory Disorders

Patients with respiratory disorders experienced significant improvements in all of the outcomes measured after rehabilitation, with the effect size in the high medium range in most cases. The improvements concerned not only lung function variables but also daily functioning, quality of life, and neuromuscular stability along with significantly lower pain perception (Table 4).

3.4 Neurological Motor and Cognitive Impairments

Clinical condition of patients with neurodegeneration-related musculoskeletal disabilities, associated with cognitive impairment, also improved significantly after rehabilitation in the health aspects investigated. The improvements were of the medium size effect, with the health-related quality of life

Table 3 Outcome measures of rehabilitative interventions in osteoarthritis patients

Outcome	Intervention – mean (SD)				Statistics			Effect size
	Pre	Post	Δ Difference	Variation	Distribution	Test	p-value	d-Cohen
FVC (L)	2.1 (0.8)	2.3 (0.9)	0.2 (0.4)	7.7%	Normal	*t*-test	0.129	0.19
FVC (% predicted)	66.7 (17.9)	70.5 (21.0)	3.8 (12.9)	5.6%	Normal	*t*-test	0.315	0.19
FEV1 (L)	2.0 (0.7)	2.2 (0.8)	0.1 (0.3)	7.0%	Normal	*t*-test	0.080	0.18
FEV1 (% predicted)	80.5 (20.6)	85.1 (22.7)	4.6 (12.2)	5.7%	Abnormal	Wilcoxon	0.162	0.21
PEF (L/min)	4.5 (1.6)	4.3 (1.6)	−0.3 (0.9)	5.8%	Abnormal	Wilcoxon	0.463	−0.17
PEF (% predicted)	68.1 (22.7)	67.0 (22.2)	−1.1 (23.1)	1.6%	Abnormal	*t*-test	0.862	−0.05
FIM (score)	109.2 (11.0)	114.3 (11.0)	5.2 (6.3)	4.7%	Normal	*t*-test	0.012	0.47
TIN (score)	17.9 (6.6)	23.5 (6.2)	5.6 (5.5)	31.4%	Abnormal	Wilcoxon	0.003	0.88
VAS (score)	6.6 (1.9)	4.6 (1.8)	−2.0 (1.5)	30.2%	Normal	*t*-test	<0.001	1.07
EQ-5D-5 L (score)	48.5 (14.6)	66.2 (11.8)	17.7 (8.1)	36.5%	Normal	*t*-test	<0.001	1.33

FVC forced vital capacity, *FEV1* forced expiratory volume in 1 s, *PEF* peak expiratory flow, *FIM* Functional Independence Measure scale, *TIN* Tinetti Gait and Balance scale, *VAS* visual analogue scale for pain perception, *EQ-5D-5 L* EuroQol health scale

Table 4 Outcome measures of rehabilitative interventions in patients with respiratory disorders

Outcome	Intervention – mean (SD)				Statistics			Effect size
	Pre	Post	Δ Difference	Variation	Distribution	Test	p-value	d-Cohen
FVC (L)	2.0 (0.8)	2.4 (0.9)	0.4 (79.4)	21.9%	Normal	*t*-test	0.045	0.51
FVC (% predicted)	66.4 (22.9)	80.4 (18.8)	14.0 (16.1)	21.1%	Normal	*t*- test	0.003	0.67
FEV1 (L)	1.6 (0.7)	1.9 (0.7)	0.3 (0.3)	20.7%	Normal	*t*- test	0.001	0.49
FEV1 (% predicted)	68.1 (25.0)	85.1 (25.1)	17.0 (20.8)	25.0%	Normal	*t*- test	0.005	0.68
PEF (L/min)	3.8 (1.9)	4.6 (1.9)	0.8 (1.4)	12.1%	Normal	*t*- test	0.038	0.43
PEF (% predicted)	60.5 (26.6)	72.7 (26.1)	12.2 (21.8)	12.0%	Abnormal	Wilcoxon	0.049	0.46
FIM (score)	100.3 (20.4)	105.7 (19.1)	5.4 (5.9)	5.4%	Abnormal	Wilcoxon	<0.001	0.28
TIN (score)	16.8 (7.8)	19.9 (6.7)	3.1 (3.6)	18.3%	Normal	*t*- test	0.004	0.63
VAS (score)	6.3 (2.6)	4.6 (2.3)	−2.8 (4.8)	27.0%	Normal	*t*- test	0.001	0.69
EQ-5D-5 L (score)	51.6 (17.2)	64.5 (15.6)	12.9 (6.5)	12.5%	Normal	*t*- test	<0.001	0.79

FVC forced vital capacity, *FEV1* forced expiratory volume in 1 s, *PEF* peak expiratory flow, *FIM* Functional Independence Measure scale, *TIN* Tinetti Gait and Balance scale, *VAS* visual analogue scale for pain perception, *EQ-5D-5 L* EuroQol health scale

Table 5 Outcome measures of rehabilitative interventions in patients with neurological motor and cognitive disorders

	Intervention – mean (SD)				Statistics			Effect size
Outcome	Pre	Post	Δ Difference	Variation	Distribution	Test	p-value	d-Cohen
FVC (% predicted)	73.0 (24.6)	81.1 (27.9)	8.1 (21.3)	11.1%	Abnormal	Wilcoxon	0.155	0.31
FEV1 (L)	2.3 (0.9)	2.5 (0.9)	0.3 (0.5)	11.4%	Normal	*t*-test	0.035	0.29
FEV1 (% predicted)	82.9 (25.1)	93.5 (30.4)	10.5 (21.4)	12.7%	Abnormal	Wilcoxon	0.020	0.38
PEF (L/min)	4.9 (1.7)	5.8 (1.8)	1.0 (1.9)	19.7%	Normal	*t*- test	0.049	0.55
PEF (% predicted)	69.3 (20.6)	84.6 (33.3)	15.4 (30.3)	22.2%	Abnormal	Wilcoxon	0.052	0.56
FIM (score)	100.8 (21.8)	107.2 (16.1)	6.5 (10.4)	6.4%	Normal	*t*- test	0.020	0.34
TIN (score)	20.0 (7.4)	23.4 (5.4)	3.4 (4.1)	17.1%	Abnormal	Wilcoxon	0.005	0.53
VAS (score)	5.4 (3.0)	3.8 (2.4)	−1.6 (2.8)	29.7%	Abnormal	Wilcoxon	0.032	0.58
EQ-5D-5 L (score)	58.2 (16.6)	73.2 (15.0)	15.0 (6.65)	25.8%	Normal	*t*- test	<0.001	0.95

FVC forced vital capacity, *FEV1* forced expiratory volume in 1 s, *PEF* peak expiratory flow, *FIM* Functional Independence Measure scale, *TIN* Tinetti Gait and Balance scale, *VAS* visual analogue scale for pain perception, *EQ-5D-5 L* EuroQol health scale

reaching the highest size effect of change in this study of about one. Rehabilitation turned out to effectively decrease perception of neuromuscular-related pain in this group of patients as well (Table 5).

4 Discussion

This study demonstrates that patients with neuromotor pathologies and with respiratory disorders benefited from the integrated bioprogressive rehabilitation approach performed in the thermal aquatic environment, the approach that focused on activation of both visceral and somatic pathways of the human body. In particular, patients regained, to some extent, autonomy in daily life functioning, their gait and balance stabilized, general quality of life improved, and perception of disease-associated pain decreased. The study confirmed the effectiveness of the integrated rehabilitation protocol in neurodegeneration-related disorders. In such patients, apart from significant improvements in the musculoskeletal system's function, lung function improved as well. Better ventilation suggests more efficient gas exchange at the lungs and thus better arterial blood oxygenation, which also includes the cortical areas of the brain and thus might be at the root of overall beneficial effects of rehabilitation in both motor and cognitive domains. Cortical blood oxygenation, fundamentally influenced by lung function, is germane to shaping responses to different kinds of sensory stimulation (Tsujii et al. 2013). Enhanced oxygenation in the cortical mantle due to midriff breathing exercise improves speech processing in dyslexia (Pecyna and Pokorski 2013).

Thermal aquatic therapies are a way to treat respiratory pathologies due to the intrinsic property of water administered as humid-hot inhalations, assisted with specific exercise (Seymour et al. 2010; Casaburi et al. 2005). In fact, Lioy et al. (1987) have reported the possibility of advantageous changes in spirometry tests in individuals subjected to humidified inhalation treatment. Treatment consisting of drinking ozonized water, due to its antioxidant effects, is also effective in reactivation of defunct innate antioxidant system, a feature of chronic inflammatory disorders (Bocci et al. 2015; Langlais et al. 1991). Positive health effects of such rehabilitative treatments might plausibly stem from mutual interaction between visceral and somatic activation. Such an interaction might be exemplified by the present observation in patients with osteoarthritis, who displayed significant improvements in mobility, gait and balance, and everyday living autonomy, associated with a reduction in pain symptoms, despite the evidently persisting degradation of cartilaginous tissue.

Somatic activation was achieved in the present study with the use of specific NMT, mudpack therapy, and hydrokinesitherapy, stimulating visceral afferent neuronal pathways that conduct impulses initiated in receptors in musculoskeletal structures. The effectiveness of NMT in association with mudpack therapy has already been demonstrated by Barassi et al. (2018b). Liu et al. (2013) have previously found benefits of mud therapy in patients with osteoarthritis. The benefits are attributable to rebalancing of afferent input to the central nervous system, which, in turn, improves motility of muscular bodily structures. Benefits of hydrokinesitherapy in neuromotor rehabilitation have been documented in multiple studies. Zamparo and Pagliaro (1998) have shown recovery of gait dynamics in the aquatic microenvironment, associated with passive mobilization to reduce spasticity in neurological disorders. In a similar vein, Barassi et al. (2019) have shown recovery of dysfunctional sensory pathways in neuromotor disorders in rehabilitation using hydrokinesitherapy. Further, Batterham et al. (2011) have shown benefits of this kind of therapy for joint mobilization and function recovery in knee and hip arthritis. The findings of the present study confirmed the benefits of hydrokinesitherapy in osteoarthritis and extended the positive effects to patients with cerebral palsy where spasticity heavily factors in gait and balance deterioration. In line with the results of Lim et al. (2005), we also found in the present study that a correction of a range of joint mobility was associated with decreased perception of pain. Symptomatic improvement is integrally associated with the patients' perception of better quality of life, the process in which psychological aspects related to pain perception play a key role (Saxena et al. 2001; Ness and Gebhart 1990).

In this study, aside from neurodegenerative disorders, lung function clearly improved in cerebral palsy patients. Cerebral palsy is a motor disorder of motion and muscles, which usually does not affect respiration per se. However, the muscles that control respiration or swallowing may be affected as well, leading to respiratory distress, upper airway infections, or aspiration pneumonia (Wilkinson et al. 2006). The integrated rehabilitative approach seems advantageous in preventing a worsening of respiratory health and in treating initially mild or recondite symptoms of a respiratory disorder before they become unmanageable. There were, however, noticeable differences in the domains of improvement, depending on the kind of disorder. Lung function failed to improve in patients suffering from osteoarthritis, although they showed a decrease in musculoskeletal and joint dysfunction, as well as in the assessment of quality of life and general health status.

In conclusion, this study investigated the plausibility of mutual interaction of sensory visceral stimulation and efferent neuronal motor pathways to musculoskeletal system in enhancing the effectiveness of medical rehabilitation. We addressed the issue by using an integrated rehabilitation protocol consisting of thermal aquatic environment, assisted by manual neuromuscular treatment, mud therapy, and hydrokinesitherapy using hot-humid inhalations in a variety of disorders underlain mostly by neurodegeneration and resulting neuromotor disability. We found in our study that this integrated rehabilitation (ICT) approach was effective in regaining patient's degree of autonomy in daily life functioning, gait and balance stabilization, and in improving health-related quality of life, at the same time decreasing pain perception. In addition, the therapy showed that activation of viscero-somatic pathways was accompanied by overlapping improvements in behavioral, affective, and cognitive aspects of patient functioning. The effectiveness of the integrated thermal aquatic rehabilitation program remains to be verified in larger patient groups of varying age, representing multifarious neuromotor and age-related disorders, and compared with traditional rehabilitation methods. Nonetheless, we believe we have shown that the integrated rehabilitation shows promise as a novel bioprogressive approach in neurological physiotherapy.

Conflict of Interest This research received no external grants or funding. The authors declare no conflicts of interest in relation to this article.

Ethical Approval All procedures performed in studies involving human participants were in accordance with the ethical standards of the institutional and/or national research committee and with the 1964 Helsinki declaration and its later amendments or comparable ethical standards.

Informed Consent Written informed consent was obtained from all individual participants included in the study.

References

Barassi G, Bellomo RG, Di Giulio C, Giannuzzo G, Irace G, Barbato C, Saggini R (2018a) Effects of manual somatic stimulation on the autonomic nervous system and posture. Adv Exp Med Biol 1070:97–109

Barassi G, Bellomo RG, Porreca A, Di Felice PA, Prosperi L, Saggini R (2018b) Somato–visceral effects in the treatment of dysmenorrhea: neuromuscular manual therapy and standard pharmacological treatment. J Altern Complement Med 24(3):291–299

Barassi G, Bellomo RG, Porreca A, Giannuzzo G, Irace G, Trivisano L, Saggini R (2019) Rehabilitation of neuromotor disabilities in aquatic microgravity environment. Adv Exp Med Biol 1113:61–73

Batterham SI, Heywood S, Keating JL (2011) Systematic review and meta–analysis comparing land and aquatic exercise for people with hip or knee arthritis on function, mobility and other health outcomes. BMC Musculoskelet Disord 12:123

Bellometti S, Cecchettin M, Galzigna L (1997) Mud pack therapy in osteoarthrosis: changes in serum levels of chondrocyte markers. Clin Chim Acta 268 (1–2):101–106

Bellomo RG, Barassi G, Di Iulio A, Giannuzzo G, Lococo A, Carmignano SM, Saggini R (2015) Rehabilitation of cancer pain in lung cancer: role of manual therapy. J Physiother Phys Rehabil 19(4)

Bellomo RG, Barassi G, Verì N, Visciano C, Giannuzzo G, Pecoraro I, Saggini R (2016) Integrated rehabilitation approach for small children with damage to the central nervous system. Adv Sci Med 1(3):5–10

Bocci V, Borrelli E, Zanardi I, Travagli V (2015) The usefulness of ozone treatment in spinal pain. Drug Des Devel Ther 9:2677

Burton AC, Taylor RM (1940) A study of the adjustment of peripheral vascular tone to the requirements of the regulation of body temperature. Am J Physiol Renal Physiol 129(3):565–577

Casaburi R, Kukafka D, Cooper CB, Witek TJ Jr, Kesten S (2005) Improvement in exercise tolerance with the combination of tiotropium and pulmonary rehabilitation in patients with COPD. Chest 127(3):809–817

Constant F, Collin JF, Guillemin F, Boulange M (1995) Effectiveness of spa therapy in chronic low back pain: a randomized clinical trial. J Rheumatol 22 (7):1315–1320

Dogaru G, Ispas A, Bulboacă A, Motricală M, Stănescu I (2017) Influence of balneal therapy at Băile Tuşnad on quality of life of post–stroke patients. Balneo Res J 8:201–205

Fioravanti A, Giannitti C, Bellisai B, Iacoponi F, Galeazzi M (2012) Efficacy of balneotherapy on pain, function and quality of life in patients with osteoarthritis of the knee. Int J Biometeorol 56(4):583–590

Grunke M, Schulze-Koops H (2006) Successful treatment of inflammatory knee osteoarthritis with tumour necrosis factor blockade. Ann Rheum Dis 65(4):555

Herdman M, Gudex C, Lloyd A, Janssen MF, Kind P, Parkin D, Badia X (2011) Development and preliminary testing of the new five–level version of EQ–5D (EQ–5D–5L). Qual Life Res 20(10):1727–1736

Holdcraft LC, Assefi N, Buchwald D (2003) Complementary and alternative medicine in fibromyalgia and related syndromes. Best Pract Res Clin Rheumatol 17 (4):667–683

Hurley MV, Walsh NE, Mitchell H, Nicholas J, Patel A (2012) Long-term outcomes and costs of an integrated rehabilitation program for chronic knee pain: a pragmatic, cluster randomized, controlled trial. Arthritis Care Res 64(2):238–247

Hutzler Y, Chacham A, Bergman U, Szeinberg A (1998) Effects of a movement and swimming program on vital capacity and water orientation skills of children with cerebral palsy. Dev Med Child Neurol 40 (3):176–181

Jochheim KA (1990) Adapted physical activity – an interdisciplinary approach. Premises, methods, and procedures. In: Adapted physical activity. Springer, Heidelberg

Konrad K, Tatrai T, Hunka A, Vereckei E, Korondi I (1992) Controlled trial of balneotherapy in treatment of low back pain. Ann Rheum Dis 51(6):820–822

Langlais B, Reckhow DA, Brink DR, Ozone in Water Treatment. Application and Engineering (1991) Cooperative research report. American Water Works Association Research Foundation. Lewis Publishers, Boca Raton/London/New York/Washington, DC

Lim HJ, Moon YI, Lee MS (2005) Effects of home–based daily exercise therapy on joint mobility, daily activity, pain, and depression in patients with ankylosing spondylitis. Rheumatol Int 25(3):225–229

Lioy PJ, Spektor D, Thurston G, Citak K, Lippmann M, Bock N, Speizer FE, Hayes C (1987) The design considerations for ozone and acid aerosol exposure and health investigations: the Fairview Lake summer camp – photochemical smog case study. Environ Int 13 (3):271–283

Liu H, Zeng C, Gao SG, Yang T, Luo W, Li YS, Lei GH (2013) The effect of mud therapy on pain relief in patients with knee osteoarthritis: a meta–analysis of randomized controlled trials. J Int Med Res 41 (5):1418–1425

Ness TJ, Gebhart GF (1990) Visceral pain: a review of experimental studies. Pain 41(2):167–234

Pappas S, Perlman A (2002) Complementary and alternative medicine: the importance of doctor–patient communication. Med Clin 86(1):1–10

Pecyna MB, Pokorski M (2013) Near–infrared hemoencephalography for monitoring blood oxygenation in prefrontal cortical areas in diagnosis and therapy of developmental dyslexia. Adv Exp Med Biol 788:175–180

Pokorski M, Barassi G, Bellomo RG, Prosperi L, Crudeli M, Saggini R (2018) Bioprogressive paradigm in physiotherapeutic and antiaging strategies: a review. Adv Exp Med Biol 1116:1–9

Pospíšil P, Konečný L, Zmeškalová M, Srovnalova H, Rektorova I, Nosavcovová E, Siegelova J (2007) Balneotherapy in patients with Parkinson's. Scr Med 80(5):233–238

Ravaud JF, Delcey M, Yelnik A (1999) Construct validity of the functional independence measure (FIM): questioning the unidimensionality of the scale and the value of FIM scores. Scand J Rehabil Med 31 (1):31–42

Restaino L, Frampton EW, Hemphill JB, Palnikar P (1995) Efficacy of ozonated water against various food–related microorganisms. J Appl Environ Microbiol 61(9):3471–3475

Saggini R, Ancona E, Carmignano SM, Supplizi M, Barassi G, Bellomo RG (2017) Effect of combined treatment with focused mechano–acoustic vibration and pharmacological therapy on bone mineral density and muscle strength in post–menopausal women. Clin Cases Miner Bone Metab 14(3):305–311

Sauvage JL, Myklebust BM, Crow-Pan J, Novak S, Millington P, Hoffman MD, Rudman D (1992) A clinical trial of strengthening and aerobic exercise to improve gait and balance in elderly male nursing home residents. Am J Phys Med Rehabil 71 (6):333–342

Saxena S, Carlson D, Billington R, Orley J (2001) The WHO quality of life assessment instrument (WHOQOL–Bref): the importance of its items for cross–cultural research. Qual Life Res 10(8):711–721

Seymour JM, Moore L, Jolley CJ, Ward K, Creasey J, Steier JS, Moxham J (2010) Outpatient pulmonary rehabilitation following acute exacerbations of COPD. Thorax 65(5):423–428

Staffieri A, Abramo (2007) A sulphurous–arsenical–ferruginous (thermal) water inhalations reduce nasal respiratory resistance and improve mucociliary clearance in patients with chronic sinonasal disease: preliminary outcomes. Acta Otolaryngol 127(6):613–617

Stein SW, Thiel CG (2017) The history of therapeutic aerosols: a chronological review. J Aerosol Med Pulm Drug Deliv 30(1):20–41

Travagli V, Zanardi I, Valacchi G, Bocci V (2010) Ozone and ozonated oils in skin diseases: a review. Mediat Inflamm 2010:610418

Travell JG, Simons DG (1983) Myofascial pain and dysfunction: the trigger point manual. Lippincott Williams Wilkins, Philadelphia

Tsujii T, Komatsu K, Sakatani K (2013) Acute effects of physical exercise on prefrontal cortex activity in older adults: a functional near–infrared spectroscopy study. In: Oxygen transport to tissue XXXIV. Springer, Berlin/Heidelberg, pp 293–298

Verhagen AP, Bierma-Zeinstra SM, Boers M, Cardoso JR, Lambeck J, de Bie RA, de Vet HC (2007) Balneotherapy for osteoarthritis. Cochrane Database Syst Rev (4):CD006864

Walsh WM, Gerley PP (1985) Thermal biofeedback and the treatment of tinnitus. Laryngoscope 95(8):987–988

Wilkinson DJ, Baikie G, Berkowitz RG, Reddihough DS (2006) Awake upper airway obstruction in children with spastic quadriplegic cerebral palsy. J Paediatr Child Health 42(1–2):44–48

Yurtkuran M, Yurtkuran MA, Dilek K, Güllülü M, Karakoc Y, Özbek L, Bingöl Ü (1999) A randomized, controlled study of balneotherapy in patients with rheumatoid arthritis. Phys Med Rehab Kuror 9 (03):92–96

Zamparo P, Pagliaro P (1998) The energy cost of level walking before and after hydro-kinesi therapy in patients with spastic paresis. Scand J Med Sci Sports 8(4):222–228

Adv Exp Med Biol - Clinical and Experimental Biomedicine (2020) 8: 39–47
https://doi.org/10.1007/5584_2019_451

Published online: 21 November 2019

Cat Allergy as a Source Intensification of Atopic Dermatitis in Adult Patients

Andrzej Kazimierz Jaworek, Krystyna Szafraniec, Magdalena Jaworek, Zbigniew Doniec, Adam Zalewski, Ryszard Kurzawa, Anna Wojas–Pelc, and Mieczyslaw Pokorski

Abstract

Atopic dermatitis (AD) is characterized by exacerbations and remissions of eczematous skin, underlain by impaired skin barrier and aberrant Th2-type and Th-22 cytokine production. A number of allergens, in particular contact with fur animals, may aggravate the disease. This study seeks to define the influence of having a regular contact with a pet cat at home on the severity of symptoms and signs of AD. We addressed the issue using the SCORing Atopic Dermatitis (SCORAD) and visual analog (VAS) scores to assess the intensity of pruritus and by measuring the blood content of specific IgE and IL-4, IL-13, and IL-22 cytokines. The study group consisted of 47 adult patients suffering from AD since childhood, 18 of whom declared having regular contact with a cat and the remaining 29 who denied it. There also was a control group consisted of 16 healthy volunteers with no AD signs. The SCORAD and VAS scores were significantly higher in patients in contact with a cat than in those without it (median SCORAD 61.0 vs. 50.4 and VAS 9.0 vs. 4.0 points, respectively). The sIgE of a majority of patients (94.4%) in contact with a cat was in Class V–VI, compared with just a few patients (3.4%) with no such contact, having sIgE in the same classes ($p < 0.001$). Significant correlations were revealed between SCORAD and VAS scores and the class level of serum sIgE value. In addition, IL-22 was a single elevated cytokine, only in the patients in contact with a cat, and it correlated with pruritus severity. The results of the study underline the need to beware of the cat fur allergen, and they stress forethought and caution in acquiring and keeping a pet cat by patients suffering from AD.

A. K. Jaworek (✉) and A. Wojas–Pelc
Department of Dermatology, Jagiellonian University Medical College, Cracow, Poland
e-mail: andrzej.jaworek@uj.edu.pl

K. Szafraniec
Department of Epidemiology and Population Studies, Institute of Public Health, Faculty of Health Sciences, Jagiellonian University Medical College, Cracow, Poland

M. Jaworek
Department of Physiotherapy, Faculty of Health Sciences, Jagiellonian University Medical College, Cracow, Poland

Z. Doniec
Department of Pneumology, Institute of Tuberculosis and Lung Disorders, Rabka, Poland

A. Zalewski
Student Scientific Group, Department of Dermatology, Jagiellonian University Medical College, Cracow, Poland

R. Kurzawa
Department of Allergology and Pneumology, Institute of Tuberculosis and Lung Disorders, Rabka, Poland

M. Pokorski
Department of Physiotherapy, Opole Medical School, Opole, Poland

Keywords
Allergy · Animal fur allergen · Atopic dermatitis · Inflammation · Interleukins · Pet cat · Skin pruritus

1 Introduction

Atopic dermatitis (AD) is a chronic inflammatory dermatosis characterized by inefficiency of the skin barrier, immunological disturbance, and dysbiosis of the saprophytic microbial flora at skin surface (Gavrilova 2018; Jaworek and Wojas-Pelc 2017). Clinically, the disorder manifests as intensive pruritus, an essential diagnostic element, which is accompanied by erythematous papules, bullae, and lichenous lesions. Skin lesions have a chronic and recurrent course, and their location depends on the patient's age. AD is considered as one of the most common skin disorders which affects about 20% of the population in the developed countries, with the number of patients in the developing countries continuously increasing as well (Weidinger and Novak 2016). According to the Polish Society of Allergology, there is approximately 1.5–2.5 million of patients with AD in Poland (Kruszewski 2012).

AD is a disease that has been known for hundreds of years, but it was distinguished as a separate medical condition not until the second half of the nineteenth century, with the pioneer works of Ernest Henri Besnier on infantile atopic dermatitis, reputed for the characterization of pruritus as the primary symptom of AD: "le premier symptome et le symptome premier" (Weyers 2013). AD, along with bronchial asthma, allergic rhinitis, conjunctivitis, and food allergy, belongs to a group of atopic disorders. Skin disease is often the first symptom of systemic atopy (Sampson et al. 2018; Weidinger and Novak 2016). Although AD is rather associated with the pediatric population as 60% of AD cases show up in the first year of life, the number of adult patients is on the rise, making the old notion of AD abating with the maturation process outdated (Leung and Guttman-Yassky 2014). Among numerous AD phenotypes, a division into extrinsic (eAD) and intrinsic (iAD) or atopiform dermatitis is widely accepted. The pathogenesis of AD consists of genetic, epigenetic, environmental, and immunological interactions, with the overlapping defect of the epidermal barrier (Nowicki et al. 2015). Classification criteria consist of the level of specific IgE (sIgE) in relation to food and airborne allergens. In eAD, particularly, Th2 lymphocyte-dependent production of cytokines (interleukins IL-4, IL-5, IL-13, and IL-31), Th17 lymphocyte-dependent (IL-17), and Th22 lymphocyte-dependent (IL-22) are considered the predominant immunological reaction (Rożalski et al. 2016). In a chronic phase of AD, Th1 lymphocytes are also involved (IL-2, IL-12, and interferon-gamma) (Gavrilova 2018; Nomura et al. 2018; Werfel et al. 2016; Mirshafiey et al. 2015).

Among the factors leading to AD exacerbations, contact with the animal fur comes to the forefront, which has been confirmed by recent studies and formulated in the expert guidelines concerning the allergen exposure withdrawal (Oncham et al. 2018; Wollenberg et al. 2018). To this end, the pet cats seem to the most frequent sensitizers and to a much greater extent than other animals. Therefore, this study seeks to define the influence of having a pet cat at home on the severity of AD symptoms. We addressed the issue using the SCORing Atopic Dermatitis (SCORAD) and visual analog (VAS) scales of pruritus severity in connection with the serum content cat sIgE. In addition, we measured the serum IL-4, IL-13, and IL-22 cytokines, modifiers of allergic inflammation.

2 Methods

2.1 Patients and Skin Lesion Assessment

The study group included 47 adult patients with AD (F/M 30/17, median age 35 years, range 19–74 years). There were 34 (72.3%) persons

Table 1 Structure of atopic dermatitis (AD) patient group ($n = 47$) and control group of healthy subjects ($n = 16$) with regard to contact with cat allergens

	Patient with AD		Control group
	Cat – yes	Cat – no	Cat – no
n	18	29	16

with mild/moderate and 13 (27.7%) persons severe AD. The control group consisted of 16 healthy volunteers (F/M 8/8, median age 29 years, range 24–67 years). Regular contact with a cat at home was defined as at least 5 days per week for 4 h a day. Structure of the study groups with regard to contact with cat allergens is shown in Table 1. In the AD group, 18 patients reported having regular contact. The remaining 29 patients denied it. The diagnosis of AD was based on the diagnostic criteria of Hanifin and Rajka (1980). The severity of skin lesions in AD was assessed using the SCORAD criteria as follows: mild <25 points, moderate AD 25–50 points, and severe >50 points (Oranje et al. 2007). Each patient carried out a self-assessment of pruritus on a VAS scale during the 24 h preceding the study as follows: no pruritus 0 points, mild pruritus 1–3 points, moderate pruritus 4–6 points, intense pruritus 7–8 points, and extremely intense pruritus 9–10 points (Schoch et al. 2017). Exclusion criteria were a lack of consent, age under 18 years, active unrelated inflammatory disorders, and a systemic anti-inflammatory therapy during the 12 months preceding the study. The patients reported the onset of skin lesions in the early childhood between the fourth month and first year of life.

2.2 Blood Biochemical Indices

Blood samples were taken from an ulnar venipuncture in a volume of 10 ml between 7:00 am and 9:00 am. They were left for 2 h at room temperature for a clot to form, centrifuged at 3,500 RPM for 10 min, and frozen at −80 °C until use. Cat sIgE (e1) was measured using a UniCAP 100 fluoro-enzymatic immunoassay (FEIA) (ImmunoCAP-System; Phadia AB, Sweden), according to manufacturer's instructions. The IgE classes were stratified as follows: Class I 0.35–0.69 IU/ml, Class II 0.70–3.49 IU/ml, Class III 3.50–17.49 IU/ml, Class IV 17.50–49.90 IU/ml, Class V 50.00–100.00 IU/ml, and Class VI >100.00 IU/ml. The cytokines IL-4, IL-13, and IL-22 were measured using an ELISA assay (R&D System, Minneapolis, MN). The detection level was set at ≥0.35 IU/ml for IgE and > 0.35 pg/ml for both sIgE and cytokines.

2.3 Statistical Elaboration

Quantitative data were expressed as medians and interquartile ranges (IQR). Differences between two groups were evaluated with the Mann-Whitney U test. For more than two groups, the Kruskal-Wallis test, followed by multiple comparison post hoc tests, was used. The strength of a relationship between quantitative data was assessed with the Spearman rank correlation coefficient. Significance of differences was assumed at the level of $\alpha \geq 0.05$. The analysis was performed using a commercial Statistica v13 package (StatSoft; Tulsa, OK).

3 Results

SCORAD score was significantly higher in the 18 patients having regular contact with a cat than that in the patients with no such contact: median 61.0, min–max 37.4–80.4, and IQR 71.3–54.1 points vs. median 50.4, min–max 10.6–75.2, and IQR 61.4–24.4 points, respectively ($p = 0.003$) (Fig. 1). Likewise, median VAS score was 9, min–max 7–10, and IQR 10–8 in the patients having regular contact with a cat vs. median 4, min–max 2–8, and IQR 6–4 points in those with no such contact ($p < 0.001$) (Fig. 2).

3.1 Specific Blood IgE

The patients with severe AD were significantly more often categorized into higher sIgE classes

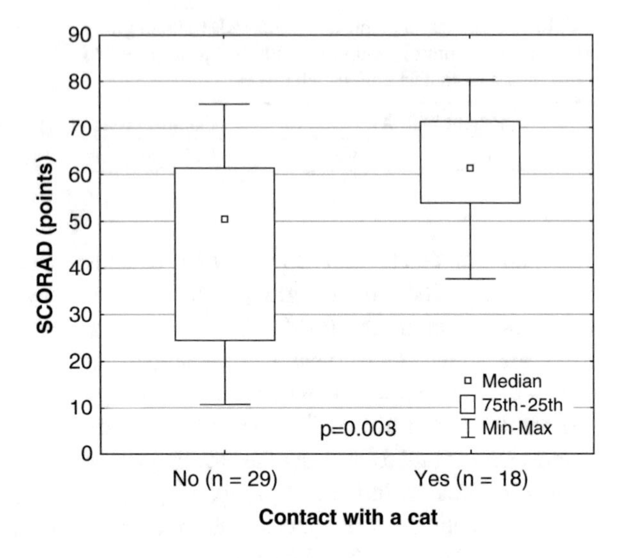

Fig. 1 Skin lesions severity in atopic dermatitis (AD) patients who declared contact with a cat (Yes) and those who did not (No), according to the SCORing Atopic Dermatitis (SCORAD) score

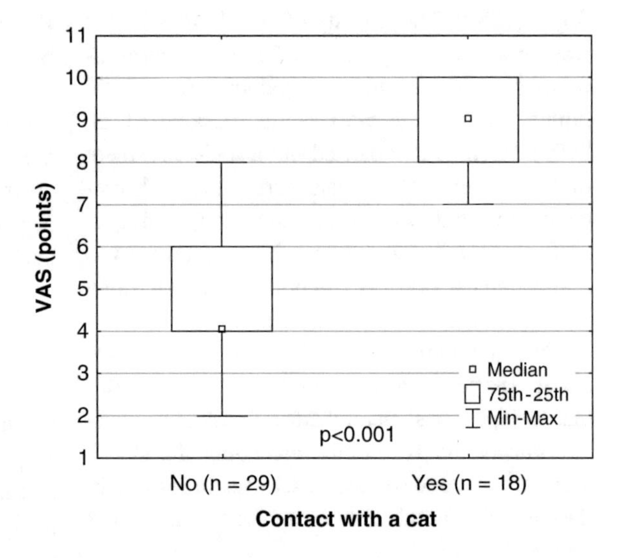

Fig. 2 Skin lesions severity in atopic dermatitis (AD) patients who declared contact with a cat (Yes) and those who did not (No), according to the visual-analog scale (VAS) score

(67.7% had Class IV–VI) than the patients with mild/moderate AD (15.4% of patients had Class IV–VI) ($p < 0.001$). Among the patients who declared having regular contact with a cat (Yes), 94.4% had sIgE in the class level V–VI, which was a significantly more numerous group of patients than those without such contact (No) of whom 3.4% had sIgE in the same class level V–VI ($p < 0.001$) (Fig. 3). Further, significant correlations were revealed between pruritus intensity in AD patients, assessed on both SCORAD and VAS scales, and sIgE class level (rho $= 0.47$, $p = 0.003$ and rho $= 0.89$, $p < 0.001$, respectively).

The control subjects were free of AD, had neither pruritic symptoms nor elevated blood sIgE levels, and thus were not included in the above presented elaboration of results concerning the AD patients who either declared or denied regular contact with a cat.

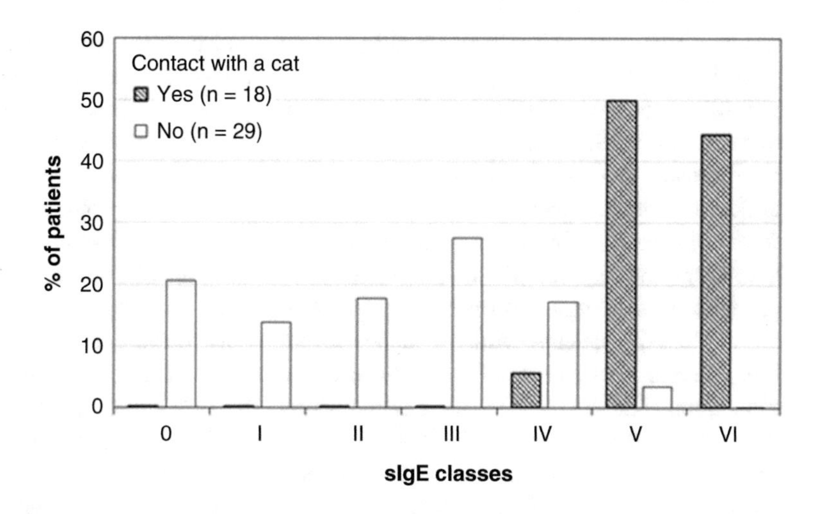

Fig. 3 Distribution of atopic dermatitis (AD) patients, who declared (Yes) or did not declare (No) contact with a cat, by sIgE classes

3.2 Blood IL-4, IL-13, and IL-22 Content

Concerning the cytokine content, IL-22 was the only one distinctly elevated in AD patients who declared regular contact with a cat. Its level was threefold greater than that in AD patients with a lack of such contact and about sixfold greater than that in the healthy subjects ($p < 0.001$). The content of IL-4 and IL-13 failed to differ appreciably among the three groups (Table 2). The analysis of a relationship between the intensity of skin lesions and IL-22 serum content across the three groups, i.e., severe AD in 34 patients vs. mild/moderate AD in 13 patients vs. healthy 16 subjects, shows that IL-22 was significantly higher in the severe AD (median 236.6 pg/ml, min–max 35.4–445.1 pg/ml) than that in mild/moderate AD (median 66.8 pg/ml, min–max 0–195.9 pg/ml), followed by that in healthy subjects (median 48.3 pg/ml, min–max 0–80.4 pg/ml) ($p < 0.001$).

IL-22 was a single cytokine that correlated with pruritus severity assessed on both SCORAD (rho = 0.52, $p < 0.001$) and VAS scales (rho = 0.60, $p < 0.001$).

4 Discussion

The significance of allergens as factors exacerbating AD is commonly accepted (Gavrilova 2018; Weidinger and Novak 2016; Werfel et al. 2016; Langan and Williams 2006). Fur animals, particularly cats, are most frequent house pets all over the world, raising concern for their role in the development of allergy. A substantial increase in sIgE we noticed in this study is consistent with other reports on the subject matter (Rożalski et al. 2016; Dhar et al. 2005). Elevated sIgG content also predisposed to intense skin

Table 2 Cytokine blood content in AD patients with and without contact with a cat and in control healthy subjects (pg/ml)

	Patients with AD				Control group		
	Cat – yes		Cat – no		Cat – no		
	Median	Min–max	Median	Min–max	Median	Min–max	*p*
IL-4	5.0	2.6–7.6	5.1	1.3–11.3	4.5	0.1–10.5	0.430
IL-13	55.2	49.8–257.9	55.2	36.3–82.2	57.9	39.0–74.1	0.860
IL-22	299.9	0.0–445.1	98.7	0.0–320.2	48.3	0.0–80.4	<0.001

lesions. Laboratory indicators of cat allergy were present in 41 out of the 47 patients investigated and were associated with the severity of pruritus, assessed on VAS. However, our patients had mostly severe AD, which ruled out the execution of skin prick tests. These findings are in line with those obtained by Mittermann et al. (2016) who have found the cat's hair the most frequent antigen raising the allergic response in a population of 179 adult patients with moderate/severe AD. Allergic reactions are often manifest in the adult population in response to environmental exposures to house pets. Oncham et al. (2018) have reported positive results of skin prick tests for cat and dog in 13% and 10% out of the 49 adult AD patients, respectively. In a Hungarian study in 34 adult AD patients, skin prick and patch tests with cat allergens, and also cat sIgE in blood serum, were positive in some of the patients. Further, in four patients of that study, exacerbation of skin lesions occurred after exposure to cats (Pónyai et al. 2008). Likewise, Will et al. (2017) have reported an unusually high level of IgE in 41 out of the 54 patients, 10 of whom were allergic to cats, with 6 of the 10 with severe AD and with 4 of the 6 who used sleeping with their beloved pets. The authors recommended lifestyle changes, consisting of withdrawal of allergen exposure, which resulted in a significant health improvement in some of the patients.

An increase in the incidence of animal allergies is ubiquitous (Asher et al. 2006). Notably, AD patients, particularly of eAD type, are predisposed to develop hypersensitivity to environmental allergens (Nomura et al. 2018; Knaysi et al. 2017). The issue of whether early contact with an animal in childhood has a protective influence on the occurrence of allergic condition in adulthood still remains unanswered (Chan and Leung 2018). A particularly interesting study concerning this issue is reported by Bisgaard et al. (2008) who have shown an interaction between a defunct gene encoding filaggrin and the development of eczema during early childhood environmental exposure, which specifically concerns cat allergens. Hon et al. (2007) have investigated 119 children, confirming the presence of allergic responses to a cat in 40 (34%) of them, using skin prick tests. Of note, only did three children declare regular contact with the animal.

In this study we also investigated the blood level of the cytokines IL-4, IL-13, and IL-22. The interleukins IL-4 and IL-13 are produced by antigen- or mitogen-induced Th2 lymphocytes, NK cells, eosinophils, innate lymphomatoid cells, mast cells, and basophils. The genes encoding for IL-4 and IL-13 are located on chromosome 5 (5q31), and the structure of both cytokines is similar. IL-4 is responsible for the class switching of antibodies produced by plasmocytes, to IgE class in case of allergy. It "recalibrates" the immune system activity to the Th2-dependent pathway, and its activation also weakens the expression of genes for filaggrin, loricrin, and involucrin which are indicators of terminal epidermal differentiation (Brandt and Sivaprasad 2011; Kim et al. 2008). IL-13's function resembles that of IL-4; receptors for both cytokines also have the same subunit. IL-13 stimulates skin inflammation, leading to fibrosis and dermis vascularization. It stimulates migration of TCD4 lymphocytes, mast cells, eosinophils, and macrophages to the dermis layer (Oh et al. 2011). Both IL-4 and IL-13 inhibit production of the antimicrobial peptides β-defensins and cathelicydins in the skin, which leads to more frequent bacterial and viral skin infections in the course of AD (Brunner et al. 2017; Chieosilapatham et al. 2017). The finding above outlined connecting IL-4 and IL-5 with inflammatory skin reactions have all been reported in adult patients with severe chronic AD (Gavrilova 2018; Nomura et al. 2018). There is a paucity of information on both cytokines concerning AD in the adult population. In this study we failed to notice any association between IL-4 or IL-13 content and clinical parameters in adults suffering from AD.

IL-22 is a cytokine which belongs to the IL-10 family and is produced mainly by T-helper 22 (Th22) lymphocytes, Th17, TCD 8 (Tc22), Tγδ lymphocytes, and others. Trifari et al. (2009)

have described a new lymphocytes population with surface receptors for CCR4, CCR6, and CCR10 chemokines, which suggests relevance to the immune system of the skin. Th22 lymphocytes produce IL-22, IL-13, tumor necrosis factor-α, and fibroblast growth factor. Lymphocytes express aryl hydrocarbon receptor (AHR) transcription factors (Nomura et al. 2018; Mirshafiey et al. 2015). The AHR ligands, such as dioxin and polycyclic aromatic hydrocarbons, promote conversion of naive TCD4+ into Th22 lymphocytes. Stimulation of naive TCD4+ by dendritic cells also promotes the development of Th22 lymphocytes. Of note, IL-22 receptors are not situated on any immune system cells but only in tissues such as the skin, kidneys, pancreas, liver, or bowels; thus they belong to tissue cytokines. IL-22 is associated with the non-specific, innate immune processes, particularly against bacteria and viruses, such as stimulation of S100 family proteins, platelet-derived growth factor A, and CXCL5 chemokines. In the skin, IL-22 stimulates keratinocyte proliferation (epidermis hyperplasia) and inhibits their terminal differentiation (Fujita 2013). Wolk et al. (2006) have shown in vitro that stimulation of keratinocytes by IL-22 induces the antimicrobial β-defensins by way of STAT3. It appears that Th22 lymphocyte performance is primarily related to the structural integrity of epidermis due to the induction of keratinocyte proliferation and migration. In case of inflammation, Th22 lymphocytes support release of cytokines and chemokines by keratinocytes (Mirshafiey et al. 2015).

Despite a consensus on the dominant role of the Th2-dependent cytokine pathway in AD, attention is also directed toward Th17-, Th1-, and especially Th22-dependent immune responses (Nomura et al. 2018; Wollenberg et al. 2018). Nograles et al. (2009) have described increased expression of IL-22 in CD4 and CD8 lymphocytes in peripheral blood in a range of skin lesions. The majority of IL-22 producing cells in skin lesions belong to the Th22 lymphocyte line, and the number of Tc22 lymphocytes in the skin associates with AD severity. There are reports showing that IL-22 content increases in both blood and skin in AD patients (Kanda and Watanabe 2012). In the blood, it associates with TARC (CCL 17), an indicator of AD severity, which in relation to CCR4 receptor expression on Th22 lymphocytes underlines the leading role of IL-22 in the pathogenesis of AD (Thijs et al. 2015; Hayashida et al. 2011). The findings of the present study showing that AD severity, assessed by SCORAD score, correlated with IL-22 blood content are in line with the role of IL-22 in AD above outlined. It should be emphasized that expression of proteins which are essential for maintaining the epidermal structure, such as filaggrin, loricrin, and involucrin, decreases in the presence of IL-22 (Brunner et al. 2016; Fujita 2013; Gutowska-Owsiak et al. 2011). It is possible that in AD patients, especially in the eAD group, high IL-22 content potentiates the defect of the epidermis barrier and, therefore, eases the exposure to allergens, which is clinically manifest by the elevated sIgE and the intensification of pruritus.

In conclusion, this study points to a key pathogenetic role of IL-22 in atopic dermatitis, particularly in the exacerbation of inflammatory skin lesions. Thus, this cytokine may become a potential target of future monoclonal therapy in atopic dermatitis. In fact, fezakinumab, a human monoclonal antibody against IL-22, is currently tested as novel treatment for atopic dermatitis, a disease otherwise highly refractory to therapy (Vakharia and Silverberg 2018).

Conflicts of Interest The authors declare no conflicts of interest in relation to this article.

Ethical Approval All procedures performed in studies involving human participants were in accordance with the ethical standards of the institutional and/or national research committee and with the 1964 Helsinki declaration and its later amendments or comparable ethical standards. The study protocol received approval of the Bioethics Committee of the Medical College of the Jagiellonian University in Cracow, Poland.

Informed Consent Written informed consent was obtained from all individual participants included in the study.

References

Asher MI, Montefort S, Björkstén B, Lai CK, Strachan DP, Weiland SK, Williams H (2006) ISAAC Phase Three Study Group. Worldwide time trends in the prevalence of symptoms of asthma, allergic rhinoconjunctivitis, and eczema in childhood: ISAAC Phases One and Three repeat multicountry cross-sectional surveys. Lancet 368(9537):733–743

Bisgaard H, Simpson A, Palmer CN, Bønnelykke K, McLean I, Mukhopadhyay S, Pipper CB, Halkjaer LB, Lipworth B, Hankinson J, Woodcock A, Custovic A (2008) Gene-environment interaction in the onset of eczema in infancy: filaggrin loss-of-function mutations enhanced by neonatal cat exposure. PLoS Med 5(6):e131

Brandt EB, Sivaprasad U (2011) Th2 cytokines and atopic dermatitis. J Clin Cell Immunol 2(3):110

Brunner PM, Khattri S, Garcet S, Finney R, Oliva M, Dutt R, Fuentes-Duculan J, Zheng X, Li X, Bonifacio KM, Kunjravia N, Coats I, Cueto I, Gilleaudeau P, Sullivan-Whalen M, Suárez-Fariñas M, Krueger JG, Guttman-Yassky E (2016) A mild topical steroid leads to progressive anti-inflammatory effects in the skin of patients with moderate-to-severe atopic dermatitis. J Allergy Clin Immunol 138(1):169–178

Brunner PM, Silverberg JI, Guttman-Yassky E, Paller AS, Kabashima K, Amagai M, Luger TA, Deleuran M, Werfel T, Eyerich K, Stingl G (2017) Increasing comorbidities suggest that atopic dermatitis is a systemic disorder. J Invest Dermatol 137(1):18–25

Chan SK, Leung DYM (2018) Dog and cat allergies: current state of diagnostic approaches and challenges. Allergy Asthma Immunol Res 10(2):97–105

Chieosilapatham P, Ogawa H, Niyonsaba F (2017) Current insights into the role of human β-defensins in atopic dermatitis. Clin Exp Immunol 190(2):155–166

Dhar S, Malakar R, Chattopadhyay S, Dhar S, Banerjee R, Ghosh A (2005) Correlation of the severity of atopic dermatitis with absolute eosinophil counts in peripheral blood and serum IgE levels. Indian J Dermatol Venereol Leprol 71(4):246–249

Fujita H (2013) The role of IL-22 and Th22 cells in human skin diseases. J Dermatol Sci 72(1):3–8

Gavrilova T (2018) Immune dysregulation in the pathogenesis of atopic dermatitis. Dermatitis 29(2):57–62

Gutowska-Owsiak D, Schaupp AL, Salimi M, Taylor S, Ogg GS (2011) Interleukin-22 downregulates filaggrin expression and affects expression of profilaggrin processing enzymes. Br J Dermatol 165(3):492–498

Hanifin JM, Rajka G (1980) Diagnostic features of atopic dermatitis. Acta Dermatol Venerol 92(Suppl):44–47

Hayashida S, Uchi H, Takeuchi S, Esaki H, Moroi Y, Furue M (2011) Significant correlation of serum IL-22 levels with CCL17 levels in atopic dermatitis. J Dermatol Sci 61(1):78–79

Hon KL, Leung TF, Lam MC, Wong KY, Chow CM, Fok TF, Ng PC (2007) Which aeroallergens are associated with eczema severity? Clin Exp Dermatol 32 (4):401–404

Jaworek AK, Wojas-Pelc A (2017) History of the atopic dermatitis – review of the selected issues since ancient times till the modern world. Przegl Dermatol 104:336–347. (in Polish)

Kanda N, Watanabe S (2012) Increased serum human β-defensin-2 levels in atopic dermatitis: relationship to IL-22 and oncostatin M. Immunobiology 217 (4):436–445

Kim BE, Leung DY, Boguniewicz M, Howell MD (2008) Loricrin and involucrin expression is down-regulated by Th2 cytokines through STAT-6. Clin Immunol 126 (3):332–337

Knaysi G, Smith AR, Wilson JM, Wisniewski JA (2017) The skin as a route of allergen exposure: part II. Allergens and role of the microbiome and environmental exposures. Curr Allergy Asthma Rep 17(1):7

Kruszewski J (2012) Definition, epidemiology and genetics of atopic dermatitis. In: Gliński W, Kruszewski J (eds) Atopic dermatitis in children and adults. Medycyna Praktyczna, Warsaw, pp 11–13. (in Polish)

Langan SM, Williams HC (2006) What causes worsening of eczema? A systematic review. Br J Dermatol 155 (3):504–514

Leung DY, Guttman-Yassky E (2014) Deciphering the complexities of atopic dermatitis: shifting paradigms in treatment approaches. J Allergy Clin Immunol 134 (4):769–779

Mirshafiey A, Simhag A, El Rouby NM, Azizi G (2015) T-helper 22 cells as a new player in chronic inflammatory skin disorders. Int J Dermatol 54(8):880–888

Mittermann I, Wikberg G, Johansson C, Lupinek C, Lundeberg L, Crameri R, Valenta R, Scheynius A (2016) IgE sensitization profiles differ between adult patients with severe and moderate atopic dermatitis. PLoS One 11(5):e0156077

Nograles KE, Zaba LC, Shemer A, Fuentes-Duculan J, Cardinale I, Kikuchi T, Ramon M, Bergman R, Krueger JG, Guttman-Yassky E (2009) IL-22-producing 'T22 T cells account for upregulated IL-22 in atopic dermatitis despite reduced IL-17-producing TH17 T cells. J Allergy Clin Immunol 123 (6):1244–1252

Nomura T, Honda T, Kabashima K (2018) Multipolarity of cytokine axes in the pathogenesis of atopic dermatitis in terms of age, race, species, disease stage and biomarkers. Int Immunol 30(9):419–428

Nowicki R, Trzeciak M, Wilkowska A, Sokołowska-Wojdyło M, Ługowska-Umer H, Barańska-Rybak W, Kaczmarski M, Kowalewski C, Kruszewski J, Maj J, Silny W, Śpiewak R, Petranyuk A (2015) Atopic dermatitis: current treatment guidelines. Statement of the experts of the Dermatological Section, Polish Society of Allergology, and the Allergology Section, Polish Society of Dermatology. Postep Dermatol Alergol 32 (4):239–249

Oh MH, Oh SY, Yu J, Myers AC, Leonard WJ, Liu YJ, Zhu Z, Zheng T (2011) IL-13 induces skin fibrosis in atopic dermatitis by thymic stromal lymphopoietin. J Immunol 186(12):7232–7242

Oncham S, Udomsubpayakul U, Laisuan W (2018) Skin prick test reactivity to aeroallergens in adult allergy clinic in Thailand: a 12-year retrospective study. Asia Pac Allergy 8(2):e17

Oranje AP, Glazenburg EJ, Wolkerstorfer A, de Waard-van der Spek FB (2007) Practical issues on interpretation of scoring atopic dermatitis: the SCORAD index, objective SCORAD and the three-item severity score. Br J Dermatol 57(4):645–648

Pónyai G, Hidvégi B, Németh I, Sas A, Temesvári E, Kárpáti S (2008) Contact and aeroallergens in adulthood atopic dermatitis. J Eur Acad Dermatol Venereol 22(11):1346–1355

Rożalski M, Rudnicka L, Samochocki Z (2016) Atopic and non-atopic eczema. Acta Dermatovenerol Croat 24 (2):110–115

Sampson HA, O'Mahony L, Burks AW, Plaut M, Lack G, Akdis CA (2018) Mechanisms of food allergy. J Allergy Clin Immunol 141(1):11–19

Schoch D, Sommer R, Augustin M, Ständer S, Blome C (2017) Patient-reported outcome measures in pruritus: a systematic review of measurement properties. J Invest Dermatol 137(10):2069–2077

Thijs J, Krastev T, Weidinger S, Buckens CF, de Bruin-Weller M, Bruijnzeel-Koomen C, Flohr C, Hijnen D (2015) Biomarkers for atopic dermatitis: a systematic review and meta-analysis. Curr Opin Allergy Clin Immunol 15(5):453–460

Trifari S, Kaplan CD, Tran EH, Crellin NK, Spits H (2009) Identification of a human helper T cell population that has abundant production of interleukin 22 and is distinct from T(H)-17, T(H)1 and T(H)2 cells. Nat Immunol 10(8):864–871

Vakharia PP, Silverberg JI (2018) New therapies for atopic dermatitis: additional treatment classes. J Am Acad Dermatol 78(3S1):76–83

Weidinger S, Novak N (2016) Atopic dermatitis. Lancet 387(10023):1109–1122

Werfel T, Allam JP, Biedermann T, Eyerich K, Gilles S, Guttman-Yassky E, Hoetzenecker W, Knol E, Simon HU, Wollenberg A, Bieber T, Lauener R, Schmid-Grendelmeier P, Traidl-Hoffmann C, Akdis CA (2016) Cellular and molecular immunologic mechanisms in patients with atopic dermatitis. J Allergy Clin Immunol 138(2):336–349

Weyers W (2013) Ernest Besnier. In: Loser C, Plewig G, Burgdorf WC (eds) Pantheon of dermatology. Springer, Berlin/Heidelberg, pp 89–94

Will BM, Severino R, Johnson DW (2017) Identification of allergens by IgE-specific testing improves outcomes in atopic dermatitis. Int J Dermatol 56(11):1150–1153

Wolk K, Witte E, Wallace E, Döcke WD, Kunz S, Asadullah K, Volk HD, Sterry W, Sabat R (2006) IL-22 regulates the expression of genes responsible for antimicrobial defense, cellular differentiation, and mobility in keratinocytes: a potential role in psoriasis. Eur J Immunol 36(5):1309–1323

Wollenberg A, Barbarot S, Bieber T, Christen-Zaech S, Deleuran M, Fink-Wagner A, Gieler U, Girolomoni G, Lau S, Muraro A, Czarnecka-Operacz M, Schäfer T, Schmid-Grendelmeier P, Simon D, Szalai Z, Szepietowski JC, Taïeb A, Torrelo A, Werfel T, Ring J (2018) Consensus-based European guidelines for treatment of atopic eczema (atopic dermatitis) in adults and children: part I. J Eur Acad Dermatol Venereol 32 (5):657–682

Adv Exp Med Biol - Clinical and Experimental Biomedicine (2020) 8: 49–56
https://doi.org/10.1007/5584_2019_438

Published online: 3 December 2019

Clinical Course and Outcome of Community-Acquired Bacterial Meningitis in Cancer Patients

Marcin Paciorek, Agnieszka Bednarska, Dominika Krogulec, Michał Makowiecki, Justyna D. Kowalska, Dominik Bursa, Anna Świderska, Joanna Puła, Joanna Raczyńska, Agata Skrzat-Klapaczyńska, Marek Radkowski, Urszula Demkow, Tomasz Laskus, and Andrzej Horban

Abstract

The aim of the study was to determine the course and outcome of bacterial meningitis (BM) in patients with cancer. We retrospectively reviewed files of patients with community-acquired BM, hospitalized in a single neuroinfection center between January 2010 and December 2017. There were 209 patients included in the analysis: 28 had cancer (9 women, 19 men; median age 76, IQR 67–80 years) and 181 were cancer-free (76 women, 105 men; median age 52, IQR 33–65 years) and constituted the control group. Cancer patients, compared with controls, were more likely to present with seizures (25% vs. 8%, $p = 0.019$), scored higher on the Sequential Organ Failure Assessment, and had a higher mortality rate (32% vs. 13%, $p = 0.025$). Further, cancer patients were less likely (64% vs. 83%, $p = 0.033$) to present with two or more out of four clinical manifestations of BM (pyrexia, neck stiffness, altered mental status, and headache) and had a lower white blood cell (WBC) count than non-cancer controls. In multiple regression analysis, the presence of bacterial meningitis in cancer patients was independently associated only with older age ($p = 0.001$) and lower WBC count ($p = 0.007$), while mortality was associated with lower Glasgow Coma Score ($p = 0.003$). In conclusion, bacterial meningitis in cancer patients is characterized by atypical symptoms and high mortality, which requires physicians' vigilance and a prompt investigation of cerebrospinal fluid in suspected cases. However, multiple regression analysis suggests that differences in clinical presentation and outcomes of bacterial meningitis between cancer and cancer-free patients may also be attributable to other factors, such as age differences.

M. Paciorek (✉), A. Bednarska, D. Krogulec, M. Makowiecki, J. D. Kowalska, D. Bursa, A. Świderska, J. Puła, J. Raczyńska, A. Skrzat-Klapaczyńska, T. Laskus, and A. Horban
Department of Adult Infectious Diseases, Warsaw Medical University, Warsaw, Poland
e-mail: mpaciorek@op.pl

M. Radkowski
Department of Immunopathology of Infectious and Parasitic Diseases, Warsaw Medical University, Warsaw, Poland

U. Demkow
Department of Laboratory Diagnostics and Clinical Immunology of Developmental Age, Warsaw Medical University, Warsaw, Poland

Keywords
Bacterial meningitis · Cancer · Community-acquired infection · Immunodeficiency · Neuroinfection

1 Introduction

Bacterial meningitis (BM) is a severe disease of the central nervous system characterized by high mortality rate and long-term neurological sequelae (Hoffman and Weber 2009); in Poland the annual incidence of BM ranged between 1.97/100,000 and 2.50/100,000 in the years 2010–2017 (Paradowska–Stankiewicz and Piotrowska 2016, 2017) which is somewhat higher than in some other European countries such as Finland (0.70/100,000) or the Netherlands (0.94/100,000) (Polkowska et al. 2017; Bijlsma et al. 2016).

The annual incidence of malignancies in Poland doubled during the last 30 years reaching 406/100,000 cases in 2013. While this number is lower than in most countries of the EU, the mortality rate in Poland is approximately 20% higher. The immune response in patients with neoplasm is commonly compromised due to the disease itself or treatment, which may increase the risk and severity of bacterial infections and may also modify their clinical course (Adriani et al. 2015; Vento and Cainelli 2003).

The aim of this study was to determine whether the presence of cancer has any influence on clinical manifestations, etiological factors, and outcome of community-acquired BM. We found that BM in patients with cancer had a somewhat different clinical presentation and worse outcome when compared with non-cancer patients.

2 Methods

In this retrospective study, we reviewed files of adult (≥ 18 years old) patients with community-acquired BM who were admitted to the Hospital for Infectious Diseases in Warsaw, Poland, between January 2010 and December 2017. The investigation was restricted to the patients diagnosed with cancer in the preceding 5 years, with exclusion of patients with brain tumors or brain metastases. Patients with community-acquired BM who had never had cancer constituted the control group. Only were the patients considered who underwent a diagnostic lumbar puncture. Those with diabetes mellitus, advanced renal disease, liver cirrhosis, alcoholism, HIV infection, transplant recipients, and patients on immunosuppressive therapy were not included in the control group. In addition, we excluded patients with meningitis secondary to head trauma, neurosurgery, and hospital-acquired infections. Clinical evaluation included scoring for Sequential Organ Failure Assessment and Glasgow Coma Scale at admission and Glasgow Outcome Score at the time of discharge.

The diagnosis of BM was based on fulfilling at least one of the following criteria: positive cerebrospinal fluid (CSF) culture, positive CSF Gram staining, typical CSF findings consisting of pleocytosis ≥100 cells/μL with ≥90% neutrophils, and a decrease in CSF glucose level below 2.2 mmol/L. Patients with CSF findings typical for BM but with negative blood and CSF cultures and negative microscopic CSF examination were considered to have BM of unknown etiology. Patients with positive blood culture, but negative CSF culture and negative CSF microscopic examination, were considered to have BM caused by a pathogen cultured from blood.

The initial antimicrobial treatment followed current Polish recommendations consisting of vancomycin and a third-generation cephalosporin in patients below 50 years of age and a combination of vancomycin with ampicillin, along with a third-generation cephalosporin, in patients older than 50 years or having cancer (Albrecht et al. 2011). The diagnosis of tuberculous meningitis was based on at least one of the following: positive culture, positive nucleic acid amplification, or positive Ehrlich-Ziehl-Neelsen staining of CSF. Tuberculous meningitis was treated with rifampicin, isoniazid, pyrazinamide, and ethambutol or streptomycin.

All patients received adjunctive therapy with dexamethasone.

Continuous data were expressed as medians with interquartile ranges (IQR) and categorical data as counts and percentages. The Mann Whitney U test was used for comparison of continuous data and the Chi-squared test for nominal variables. Multiple logistic regression analysis with adjusted odds ratios (OR) and confidence intervals (CI) was used to determine variables independently associated with the presence of cancer and independently influencing mortality. A p-value of <0.05 defined statistically significant differences. The analysis was performed using the statistical package R v3.5.2 available at https://www.r-project.org.

3 Results

Overall, 340 patients were hospitalized with community-acquired BM in the years 2010–2017, of whom 122 were excluded from the evaluation due to comorbidities such as diabetes mellitus in 67 patients, alcoholism in 45, immunosuppressive therapy in 41, liver cirrhosis in 10, and organ transplantation in 1 patient. We also excluded nine patients with primary brain tumors or brain metastases. Final evaluation included 209 patients, of whom 28 had cancer (9 women and 19 men; median age 76, IQR 67–80 years). Types of cancer in patients with BM are displayed in Table 1. In addition, there were 181 non-cancer patients (76 women and 105 men; median age 52, IQR 33–65 years) who were free of comorbidities that could potentially adversely affect the immune system. The non-cancer patients served as the control group for this study.

Cancer patients were significantly older ($p < 0.001$) than non-cancer controls. At admission, cancer patients were more likely to present with seizures when compared with controls (25% vs. 8%; $p = 0.019$, respectively) (Table 2), but they were less likely to present with at least two of the four classical manifestations of BM such as pyrexia, neck stiffness, altered mental status, and headache (64% vs. 83%; $p = 0.033$, respectively). Furthermore, they scored higher in the Sequential Organ Failure Assessment when compared with controls (median 4, IQR 1–6 vs. median 2, IQR 0–4; $p = 0.046$, respectively), and their mortality was higher (32% vs. 13%; $p = 0.025$, respectively). Clinical outcome reflected by the Glasgow Outcome Score tended to be worse in cancer patients than in controls (median 4, IQR 1–5 vs. median 5, IQR 3–5; $p > 0.05$, respectively).

The etiological agent was identified only in 68% of cancer patients and in 60% of non-cancer controls (Table 2). There were no significant differences between the two groups: *Streptococcus pneumoniae* and *Staphylococci*, followed by *Listeria monocytogenes* and *Mycobacterium tuberculosis*, were the most common pathogens in cancer patients, while *Streptococcus pneumoniae*, followed by *Staphylococci* and *Neisseria meningitides*, were in non-cancerous patients.

The results of blood or CSF indices in cancer and non-cancer patients were grossly akin to one another, except the white blood count that was

Table 1 Cancer type in 28 patients with bacterial meningitis

Solid tumors 13/28 (52%)	Hematologic malignancies 15/28 (48%)
Colon – 3	Chronic lymphoid leukemia – 5
Urinary tract – 2	Multiple myeloma – 4
Skin 2	Non-Hodgkin lymphoma – 3
Stomach – 1	Diffuse large B-cell lymphoma – 1
Gallbladder – 1	Hodgkin lymphoma – 1
Lung – 1	Acute lymphoid leukemia – 1
Liver – 1	
Nasopharyngeal – 1	
Salivary gland – 1	

Data are presented as counts (%)

Table 2 Demographic, clinical, and etiological admission data in bacterial meningitis patients with and without cancer

Characteristics of patients	Cancer ($n = 28$)	Non-cancer ($n = 181$)	*p*-value
Age (year)	76 (67–80)	52 (33–65)	**<0.001**
Male gender	19/28 (68%)	105/181 (58%)	0.422
Pyrexia	19/27 (70%)	155/179 (87%)	0.060
Neck stiffness	20/27 (74%)	138/176 (78%)	0.798
Altered mental status	5/28 (18%)	29/179 (16%)	1.000
Triad of symptoms*	3/28 (11%)	20/181 (11%)	0.998
Headache	11/26 (42%)	108/176 (61%)	0.103
Two out of four symptoms**	18/28 (64%)	151/181 (83%)	**0.033**
Nausea and/or vomiting	8/28 (29%)	67/178 (38%)	0.474
Rush	1/28 (4%)	20/177 (11%)	0.359
Seizures	7/28 (25%)	15/180 (8%)	**0.019**
Vertebral pain	3/27 (11%)	16/176 (9%)	1.000
Ataxia, dysfunction of motor coordination, or balance disorders	2/27 (7%)	10/178 (6%)	1.000
Speech dysfunction (aphasia, dysarthria)	1/28 (4%)	16/177 (9%)	0.201
Cranial nerve palsies	2/28 (7%)	11/180 (6%)	1.000
Paresis/plegia	0/28 (0%)	16/180 (9%)	0.207
Hearing disorders	2/28 (7%)	19/176 (11%)	0.394
Memory disorders	1/27 (4%)	6/178 (3%)	1.000
Disease severity/outcome			
Glasgow coma scale	12 (8–14)	13 (8–15)	0.612
SOFA score on admission	4 (1–6)	2 (0–4)	**0.046**
ICU admission (%)	13/28 (46%)	62/181 (34%)	0.299
Glasgow outcome score	4 (1–5)	5 (3–5)	0.122
Mortality (%)	9/28 (32%)	24/181 (13%)	**0.025**
Identified pathogen (%)			
Streptococcus pneumoniae	5/28 (18%)	36/181 (20%)	1.000
Neisseria meningitidis	1/28 (4%)	18/181 (10%)	0.460
Listeria monocytogenes	3/28 (11%)	7/181 (4%)	0.270
Staphylococcus	5/28 (18%)	21/181 (12%)	0.532
Haemophilus influenzae	0/28 (0%)	3/181 (2%)	1.000
Other gram-positive bacteria	0/28 (0%)	14/181 (8%)	0.264
Other gram-negative bacteria	2/28 (7%)	3/181 (2%)	0.270
Mycobacterium tuberculosis	3/28 (11%)	6/181 (3%)	0.195
Etiology unknown	9/28 (32%)	73/181 (40%)	1.000

Data are counts (%) or medians (IQR); percentage values are round off to the nearest unit; *SOFA* Sequential Organ Failure Assessment, *ICU* intensive care unit; *pyrexia, neck stiffness, and altered mental status; **presence of at least two out of the four symptoms: pyrexia, neck stiffness, altered mental status, and headache. Significant differences between cancer and non-cancer patients are marked in bold

lower in the former group (median 10.4, IQR 5.9–15.5 vs. median 16, IQR 10.7–20.6 G/L; $p = 0.016$) (Table 3).

However, in multiple logistic regression analysis (Table 4), the presence of cancer was independently associated only with a lower risk of WBC count elevation (OR 0.89, 95% CI 0.82–0.964; $p = 0.007$) and with older age (OR 1.06, 95% CI 1.03–1.10; $p = 0.001$). Furthermore, the Glasgow Coma Scale was the only factor independently associated with mortality (Table 5).

4 Discussion

In this study, the prevalence of cancer in patients with community-acquired BM was 10.9%, which

Table 3 Laboratory tests in bacterial meningitis patients with and without cancer

Blood tests	Cancer	Non-cancer	*p*-value
CRP (mg/L)	90 (36–278)	223 (78–328)	0.112
Lactic acid (mmol/L)	1.7 (1.4–2.7)	2.0 (1.6–2.8)	0.568
WBC (G/L)	10.4 (5.9–15.5)	16.00 (10.7–20.6)	**0.016**
PLT (G/L)	140 (102–229)	193 (100–239)	0.279
PCT (ng/mL)	1.2 (0.1–7.0)	5.0 (0.5–15.5)	0.439
D-dimers µg/L	2,240 (1,266–3,622)	2,114 (1,093–3,879)	0.136
Creatinine µmol/L	76.5 (61.5–116.0)	68.0 (56.0–82.0)	0.129
Urea concentration (mmol/L)	8.5 (5.6–13.3)	5.6 (4.0–8.0)	0.059
Cerebrospinal fluid tests			
Cytosis (cells/µL)	872 (219–2,057)	1,510 (312–5,635)	0.413
Granulocytes (%)	72.5 (40.0–87.8)	89.0 (73.0–95.0)	0.082
Protein (g/L)	3.6 (1.6–6.8)	2.8 (1.3–5.7)	0.391
Glucose (mmol/L)	1.9 (1.1–3.9)	1.8 (0.0–3.0)	0.329
Lactic acid (mmol/L)	5.3 (3.8–9.9)	5.2 (3.3–10.6)	0.434
Chlorides (mmol/L)	118.0 (112.3–121.0)	117.0 (113.0–121.0)	0.977

Data are presented as medians (IQR). *CRP* C-reactive protein, *WBC* white blood cells, *PLT* platelets, *PCT* procalcitonin, *CSF* cerebrospinal fluid. Significant difference between cancer and non-cancer patients is marked in bold

Table 4 Multiple logistic regression analysis of factors independently associated with the presence of cancer in patients with bacterial meningitis

Variable	OR	95% CI	*p*-value
Age	1.06	1.03–1.10	**0.001**
Two out of four symptoms*	0.56	0.17–1.98	0.352
SOFA	0.94	0.78–1.13	0.537
WBC	0.89	0.82–0.96	**0.007**
Mortality	2.74	0.74–11.39	0.158
Seizures	4.15	0.66–11.19	0.081

Data are odds ratio (OR) and 95% confidence interval (95% CI). *SOFA* sepsis-related Sequential Organ Failure Assessment score; *presence of at least two out of four symptoms: pyrexia, neck stiffness, altered mental status, and headache; *WBC* white blood cell count. Significant differences between cancer and non-cancer patients are marked in bold

Table 5 Multiple logistic regression analysis of factors independently associated with mortality in patients with bacterial meningitis

Variable	OR	95% CI	*p*-value
Age	1.00	0.97–1.04	0.861
GCS	0.72	0.57–0.89	**0.003**
SOFA	1.05	0.80–1.36	0.743
Urea concentration	1.13	1.00–1.30	0.070
Two out of four symptoms*	0.40	0.09–1.59	0.200
Streptococcus pneumoniae	1.56	0.38–6.24	0.526
Cancer	2.29	0.48–11.15	0.294

Data are odds ratio (OR) and 95% confidence interval (95% CI). *GCS* Glasgow Coma Scale; *SOFA* sepsis-related Sequential Organ Failure Assessment score; *presence of at least two out of four symptoms: pyrexia, neck stiffness, altered mental status, and headache. Significant difference between cancer and non-cancer patients is marked in bold is in line with the findings reported by others. In a study of Pomar et al. (2017), which investigated a cohort of BM patients from a single center in Spain, the prevalence of cancer was 15%, while in a nationwide study of Costerus et al. (2016) in the Netherlands, it was 11%. Furthermore, in the latter study, the annual incidence of BM was estimated to be from 2.7- to 3.5-fold greater in patients with cancer. The reasons for a greater susceptibility and incidence of BM in cancer patients, particularly those with hematologic malignancies, are multifactorial and include bone marrow invasion, organ dysfunction caused by narrowing of natural pathways, destruction of anatomical barriers, chemotherapy-induced granulocytopenia, and malnutrition (Pruitt 2012; Maschmeyer and Haas 2008; Guven et al. 2006).

The present study revealed a number of differences in the clinical presentation between cancer and non-cancer patients. The presence of at least two out of the four common manifestations of BM (pyrexia, neck stiffness, altered mental status, and headache) was less frequent in the former group (64% vs. 83%; $p = 0.033$). These four symptoms are common clinical manifestations of BM in adults (van de Beek et al. 2016). For instance, in a study of van de Beek et al. (2004), 95% of patients with BM presented with at least two out of these symptoms. The present results are compatible with an earlier study of Pomar et al. (2017) who found fever, neck stiffness, and headache to be less frequent in cancer patients. A scant presentation of typical symptoms may delay the diagnosis and therapy initiation (Domingo et al. 2013).

Another difference in the clinical manifestation of BM in cancer and non-cancer patients was a greater incidence of seizures (25% vs. 8%; $p = 0.019$) in the former, despite that patients with brain tumors or metastases were excluded from the study. This finding is at variance with those of former studies where the incidence of seizures in cancer patients was 8.2% and 19% (Pomar et al. 2017; Costerus et al. 2016), and it was akin to that found in non-cancer patients. One possibility for the discrepancy is exclusion of chronic alcohol abusers from the control group in the present study. Alcohol abusers constitute a significant percentage of BM as they are predisposed to bacterial infections and seizures which are common manifestations of alcohol withdrawal (van Veen et al. 2017; Weisfelt et al. 2010). It has been previously reported that in patients with BM, seizures are associated with poor outcome, attributable to increased intracranial pressure, lactic acidosis, and comorbidities such as immunocompromised state, pneumoniae, and hydrocephalus (Zoons et al. 2008; Wang et al. 2005).

In this study, routine blood and CSF laboratory tests did not reveal any significant differences with the exception of peripheral blood leukocyte count being lower in patients with cancer; the difference persisted in multiple logistic regression analysis. These findings are compatible with two previous reports analyzing the cancer patients with BM (Pomar et al. 2017; Costerus et al. 2016). A lower WBC in cancer patients with BM may be a result of metastases into the bone marrow, impaired hematopoiesis caused by chemotherapy, radiation, or the expression of a suppressive effect of anti-inflammatory mediators, such as IL-10 (Zhao et al. 2015). We also found that patients with cancer scored higher on the SOFA scale at admission. This scale, a crucial part of current sepsis definition (Singer et al. 2016), assesses multiple organ dysfunction, and it correlates with outcome. In some previous studies, a higher SOFA score, along with lower GCS (Pietraszek–Grzywaczewska et al. 2016), was found a predictor of an unfavorable outcome in patients with BM.

In this study, patients with cancer were significantly older than non-cancer patients, which was not unexpected as the prevalence of cancer increases with age (Bellizzi and Gosley 2012; Harding et al. 2012). Of note, it has been reported that elderly patients with BM are more likely to present with atypical clinical manifestation, are predisposed to different etiological factors, and have a higher mortality rate (Cabellos et al. 2009), which raises the possibility that the observed differences among our patients were largely age-related. Indeed, in multiple logistic regression analysis, the presence of cancer was associated with older age and lower WBC only. A mortality rate among our cancer patients was significantly higher than that among controls (32% vs. 13%, respectively; $p = 0.025$). This is in line with a previous study of Pomar et al. (2017) in which the overall mortality of patients with cancer was significantly higher compared with patients without malignancy (31% vs. 16%, respectively). However, in a multivariate analysis, mortality in our patients was associated with the Glasgow Coma Scale rather than with the presence of cancer.

Streptococcus pneumoniae and *Staphylococci* were the most common causative agents in patients of this study, and no significant differences were found between cancer and non-cancer patients concerning the etiological factors involved. *Listeria monocytogenes* also was a common etiological agent, and it accounted for 11% of cases. However, the prevalence of this bacteria was lower when compared with the 21% reported in the studies of Costerus et al. (2016) and Pomar et al. (2017). The presence of malignancy was identified as a risk factor for listeriosis in a large epidemiological investigation of Mook et al. (2011) who have reported that patients with cancer have a fivefold greater risk for this infection. While we found *Listeria* to be more frequent in patients with cancer, the difference failed to reach statistical significance. The shortcomings of the present study are that we did not differentiate between patients with active cancer and those in therapy-induced remission nor did we differentiate between patients with solid tumors and hematological malignancies, all of whom were evaluated as a single group.

In conclusion, it is advisable to be vigilant for the presence of bacterial meningitis in patients with cancer as clinical signs and symptoms are often atypical, while the overall mortality is high. However, since the presence of cancer was independently associated only with older age, the observed differences between patients with bacterial meningitis on the background of cancer and without cancer appear likely to be of multifactorial origin.

Acknowledgments MP was supported by grant FRN 004/2019 from the Research Development Foundation of the Hospital for Infectious Diseases in Warsaw and TL by grant 2017/25/B/NZ6/01463 from the National Science Center.

Ethical Approval All procedures performed in studies involving human participants were in accordance with the ethical standards of the institutional and/or national research committee and with the 1964 Helsinki declaration and its later amendments or comparable ethical standards.

Informed Consent This study had a retrospective nature consisting of reviewing medical files, with no direct contact with patients; therefore, the requirement of obtaining individual patient consent did not apply.

References

Adriani KS, Brouwer MC, van de Beek D (2015) Risk factors for community-acquired bacterial meningitis in adults. Neth J Med 73(2):53–60

Albrecht P, Hryniewicz W, Kuch A, Przyjałkowski W, Skoczyńska A, Szenborn L (2011) Recommendations in bacterial infections of the central nervous system. https://koroun.edu.pl/wp-content/uploads/2017/10/Rekomendacje-ukl-nerwowy_2011.pdf. Accessed on 25 August 2019 (Article in Polish)

Bellizzi KM, Gosley MA (2012) Cancer and aging handbook: research and practice, 1st edn. Wiley–Blackwell, Hoboken

Bijlsma MW, Brouwer MC, Kasanmoentalib ES, Kloek AT, Lucas MJ, Tanck MW, van der Ende A, van de Beek D (2016) Community-acquired bacterial meningitis in adults in the Netherlands, 2006–14: a prospective cohort study. Lancet Infect Dis 16(3):339–347

Cabellos C, Verdaguer R, Olmo M, Fernandez-Sabe N, Cisnal M, Ariza J, Gudiol F, Viladrich PF (2009) Community-acquired bacterial meningitis in elderly patients: experience over 30 years. Medicine (Baltimore) 88(2):115–119

Costerus JM, Brouwer MC, van der Ende A, van de Beek D (2016) Community-acquired bacterial meningitis in adults with cancer or a history of cancer. Neurology 86 (9):860–866

Domingo P, Pomar V, Benito N, Coll P (2013) The changing pattern of bacterial meningitis in adult patients at a large tertiary university hospital in Barcelona, Spain (1982–2010). J Infect 66(2):147–154

Guven GS, Uzun O, Cakir B, Akova M, Unal S (2006) Infectious complications in patients with hematological malignancies consulted by the infectious diseases team: a retrospective cohort study (1997–2001). Support Care Cancer 14(1):52–55

Harding C, Pompei F, Wilson R (2012) Peak and decline in cancer incidence, mortality, and prevalence at old ages. Cancer 118(5):1371–1386

Hoffman O, Weber RJ (2009) Pathophysiology and treatment of bacterial meningitis. Ther Adv Neurol Disord 2(6):1–7

Maschmeyer G, Haas A (2008) The epidemiology and treatment of infections in cancer patients. Int J Antimicrob Agents 31(3):193–197

Mook P, O'Brien SJ, Gillespie IA (2011) Concurrent conditions and human listeriosis, England, 1999–2009. Emerg Infect Dis 17(1):38–43

Paradowska–Stankiewicz I, Piotrowska A (2016) Meningitis and encephalitis in Poland in 2014. Przegl Epidemiol 70(3):349–357

Paradowska–Stankiewicz I, Piotrowska A (2017) Meningitis and encephalitis in Poland in 2015. Przegl Epidemiol 71(4):493–500

Pietraszek–Grzywaczewska I, Bernas S, Lojko P, Piechota A, Piechota M (2016) Predictive value of the APACHE II, SAPS II, SOFA and GCS scoring systems in patients with severe purulent bacterial meningitis. Anaesthesiol Intensive Ther 48(3):175–179

Polkowska A, Toropainen M, Ollgren J, Lyytikainen O, Nuorti JP (2017) Bacterial meningitis in Finland, 1995–2014: a population–based observational study. BMJ Open 7(5):e015080

Pomar V, Benito N, Lopez–Contreras J, Coll P, Gurgui M, Domingo P (2017) Characteristics and outcome of spontaneous bacterial meningitis in patients with cancer compared with patients without cancer. Medicine (Baltimore) 96(19):e6899

Pruitt AA (2012) CNS infections in patients with cancer. Continuum (Minneap Minn) 18(2):384–405

Singer M, Deutschman CS, Seymour CW, Shankar–Hari M, Annane D, Bauer M, Bellomo R, Bernard GR, Chiche JD, Coopersmith CM, Hotchkiss RS, Levy MM, Marshall JC, Martin GS, Opal SM, Rubenfeld GD, van der Poll T, Vincent JL, Angus DC (2016) The third international consensus definitions for Sepsis and septic shock (Sepsis–3). JAMA 315(8):801–810

Van de Beek D, de Gans J, Spanjaard L, Weisfelt M, Reitsma JB, Vermeulen M (2004) Clinical features and prognostic factors in adults with bacterial meningitis. N Engl J Med 351(18):1849–1859

Van de Beek D, Cabellos C, Dzupova O, Esposito S, Klein M, Kloek AT, Leib SL, Mourvillier B, Ostergaard C, Pagliano P, Pfister HW, Read RC, Sipahi OR, Brouwer MC (2016) ESCMID guideline: diagnosis and treatment of acute bacterial meningitis. Clin Microbiol Infect 22:S37–S62

van Veen KE, Brouwer MC, van der Ende A, van de Beek D (2017) Bacterial meningitis in alcoholic patients: a population-based prospective study. J Infect 74 (4):352–357

Vento S, Cainelli F (2003) Infections in patients with cancer undergoing chemotherapy: aetiology, prevention, and treatment. Lancet Oncol 4(10):595–604

Wang KW, Chang WN, Chang HW, Chuang YC, Tsai NW, Wang HC, Lu CH (2005) The significance of seizures and other predictive factors during the acute illness for the long-term outcome after bacterial meningitis. Seizure 14(8):586–592

Weisfelt M, de Gans J, van der Ende A, van de Beek D (2010) Community-acquired bacterial meningitis in alcoholic patients. PLoS One 5(2):e9102

Zhao S, Wu D, Wu P, Wang Z, Huang J (2015) Serum IL-10 predicts worse outcome in cancer patients: a meta-analysis. PLoS One 10(10):e0139598

Zoons E, Weisfelt M, de Gans J, Spanjaard L, Koelman JH, Reitsma JB, van de Beek D (2008) Seizures in adults with bacterial meningitis. Neurology 70(22 Pt 2):2109–2115

Adv Exp Med Biol - Clinical and Experimental Biomedicine (2020) 8: 57–70
https://doi.org/10.1007/5584_2019_448

Published online: 5 December 2019

Influence of Coping Strategy on Perception of Anxiety and Depression in Patients with Non-small Cell Lung Cancer

Beata Jankowska-Polańska, Jacek Polański,
Mariusz Chabowski, Joanna Rosińczuk, and Grzegorz Mazur

Abstract

The purpose of this study was to evaluate the influence of cognitive adjustment to cancer, assessed on the mini-Mental Adjustment to Cancer (mini-MAC) scale, on perception of anxiety and depression, assessed with the Hospital Anxiety and Depression Scale, in patients with non-small cell lung carcinoma (NSCLC). There were 185 patients, grouped according to the score of mini-MAC into constructive coping strategies, balanced coping strategies, and destructive coping strategies. We found that patients with predominantly destructive coping strategies had a higher level of anxiety than those with balanced or constructive strategies (10.9 vs. 9.3 vs. 6.3 points, respectively; $p < 0.001$). Likewise, symptoms of depression were more pronounced in patients having destructive coping strategies than in those with balanced or constructive strategies (11.9 vs. 8.8 vs. 5.8 points, respectively; $p < 0.001$). We further found that constructive coping strategy was a significant independent predictor of lower levels of anxiety and depressive symptoms. Other predictors included symptomatic treatment and a good nutritional status, while pain, chemotherapy, and poor performance status exacerbated the negative emotions. We conclude that cognitive adjustments to having cancer outstandingly modify the development of anxiety and depression in NSCLC patients, which also influences the choice of treatment and the treatment process itself. Thus, psychological assessment is essential in clinical practice and care for patients with lung cancer.

Keywords

Anxiety · Cognitive adjustment · Coping strategy · Depression · NSCLC

B. Jankowska-Polańska (✉)
Department of Clinical Nursing, Faculty of Health Sciences, Wroclaw Medical University, Wroclaw, Poland
e-mail: beata.jankowska-polanska@umed.wroc.pl

J. Polański
Department of Internal Medicine, Occupational Diseases, Hypertension and Clinical Oncology, Wroclaw Medical University, Wroclaw, Poland

M. Chabowski
Division of Surgical Procedures, Department of Clinical Nursing, Faculty of Health Sciences, Wroclaw Medical University, Wroclaw, Poland

Department of Surgery, Fourth Military Teaching Hospital, Wroclaw, Poland

J. Rosińczuk
Department of Nervous System Diseases, Faculty of Health Sciences, Wroclaw Medical University, Wroclaw, Poland

G. Mazur
Department of Internal Medicine, Occupational Diseases, Hypertension and Clinical Oncology, Wroclaw Medical University, Wroclaw, Poland

1 Introduction

Lung cancer is the most common cancer and the leading cause of cancer-related death globally and its incidence is expected to continue growing in the future (Parkin et al. 1999). The majority of lung cancers (80–85%) are classified as non-small cell lung carcinoma (NSCLC). Patients usually have advanced disease at diagnosis. Thus, prognosis is often poor, with chemotherapy only improving quality of life and a 1-year survival (Bray et al. 2018; Grønberg et al. 2010).

Patients with lung cancer have a considerably higher rate of psychological distress when compared to other types of cancer (Carlson et al. 2019; Zabora et al. 2001). They also are at a greater risk of psycho-social problems throughout the treatment process. Anxiety affects 16–49% and depression affects 23–40% of lung cancer patients (Kim et al. 2005; Gonzalez and Jacobsen 2012). The disparity of the findings outlined above may result from differences in the methods and protocols of investigations, and from patient-related factors such as age, cancer stage, or performance status (Iyer et al. 2013; Hopwood and Stephens 2000). Over the last decade, a holistic approach to cancer treatment has been gaining recognition, including an assessment of emotional problems associated with the disease and its treatment, apart from the classical measures of symptom severity, survival, and satisfaction with treatment. Cancer diagnosis and treatment cause a deterioration in physical performance, interfere with social and family life, and produce mental discomfort typically manifesting as depression or anxiety (Shimizu et al. 2015; Arrieta et al. 2013; Carlsen et al. 2005; Faller and Schmidt 2004).

Anxiety is an adaptive response that motivates the patient to comply with the proposed diagnostic and therapeutic interventions. However, it may also become a clinically poorly acceptable sense of distress that may be a cause of, or be associated with, negative outcomes such as disturbances in patient's daily functioning (House and Stark 2002), less effective medical decision making (Latini et al. 2007), exacerbation of medical symptoms (Pirl et al. 2012; Montazeri et al. 2001), disruptions in cancer care (Greer et al. 2008), and a poorer quality of life (Polański et al. 2018; Brown et al. 2010). Patients' reaction to chronic illness vary depending on multiple factors, including the possibility of combating the disease, personality, acceptance of illness, potential complications, and an increasingly discussed sense of coherence (Sak et al. 2012). The adaptation to limitations imposed by cancer depend, to a great extent, on the patient's strategies for coping with difficult and stressful situations. Lazarus and Folkman (1984) have described the coping as "constantly changing cognitive and behavioral efforts to manage specific external or internal demands that are appraised as taxing or exceeding the resources of a person". According to this definition, perception of one's situation, when faced with a disease, has three components: threat, harm, and challenge. These three components are the basis to develop a specific strategy for coping with disease. A related theory of mental adjustment to cancer, developed by Watson et al. (1988), defines the adjustment as "the cognitive and behavioral responses the patient makes to the diagnosis of cancer". There is a basic difference between the two theories as in the former the emotional response results from the coping strategy and the latter includes the emotional responses to threatening events. In this article, the mental adjustment theory expressed as the coping strategy includes the emotional response.

The scientific literature increasingly includes the assessment of a patient's subjective view of an illness in predicting outcomes of chronic treatment. The assessment of patient's attitude usually focuses on the strategy to cope with stress, and the associations with symptom severity and quality of life. Patients who use effective methods of coping are able to maintain an active, positive attitude toward disease and treatment, which helps decrease symptoms. Contrarily, poor coping may exacerbate symptoms and distress, adversely affecting the physical and emotional condition (Chabowski et al. 2018a; Vaillo et al. 2018; Mosher et al. 2015). Mental adjustment and coping are described as factors affecting patients' psychological state and perceived quality of life.

Among specific responses in cancer patients, "a fighting spirit" is described as a highly optimistic response, and "information seeking" as a beneficial response leading, for instance, to fewer recurrences or metastases, while "helplessness/hopelessness" is seen as a poor response adversely affecting psychological health and perceived quality of life, shortening 5- and 10-year survivals (Czerw et al. 2015; Fitzell and Pakenham 2010). There are, however, reports that question the potential influence of a cognitive reaction to the appearance of a cancer disease on health outcomes (Arrieta et al. 2013; Genç and Tan 2011; Prasertsri et al. 2011). Due to the paucity of data and controversy surrounding the issue of cognitive adjustment and coping in cancer patients, we set out in this study to define the intensity of anxiety and depression symptoms in lung cancer patients, depending on the strategy for coping with disease, and to identify the socio-clinical determinants that could lead to negative emotions arising in patients.

2 Methods

2.1 Patients and Questionnaires

There were 185 patients (mean age 62.7 ± 9.7 years) included in this survey study, with a clinically confirmed diagnosis of NSCLC. The patients were treated at the Lower Silesian Lung Center in Poland. They were grouped according to the score of the mini-Mental Adjustment to Cancer (mini-MAC) scale that we used into constructive coping strategies ($n = 41$), balanced coping strategies ($n = 56$), and destructive coping strategies ($n = 88$). Inclusion criteria were as follows: age >18 years; consent to participate; and comprehending the survey questions. Exclusion criteria were as follows: uncertain cancer diagnosis; coexistence of other severe chronic diseases that could interfere with the patient's perception of health status such as other malignant tumors, mental disorders, and other conditions that could make the patient unfit to fill in surveys. Each patient underwent clinical examinations, chest X-ray, pulmonary function tests, and biochemical blood tests along with blood gas content analysis. Clinical data were retrieved from the patients' medical files.

The mini-MAC scale is a self-reported psychometric tool, developed by Watson et al. (1994) and modified for the Polish population by Juczyński (2001). It consists of 29 items that assess four coping strategies: 1. anxious preoccupation (anxiety about the disease and seeing the disease as something alarming, uncontrollable, and threatening); 2. fighting spirit (seeing the disease as a challenge, with active efforts to seek complementary therapies, often including dancing, traveling, or exercising); 3. helplessness/hopelessness (a sense of confusion and helplessness, often entailing a withdrawal from activities, and giving up work); and 4. positive redefinition (changing one's attitude toward life and appreciation of its value in the face of the disease). Anxious preoccupation and helplessness/hopelessness are components of the passive or destructive coping strategy, while the other two strategies are part of the active or constructive coping strategy. Each statement in the mini-MAC questionnaire is rated at a four-point scale with "1" denoting definitely disagree and "4" denoting definitely agree. The score for each coping strategy is calculated separately by adding scores from specific items; it ranges between 7 and 28 points. The higher the score, the greater is the intensity of behaviors associated with a given coping strategy.

We also used the Hospital Anxiety and Depression Scale (HADS) that consists of 14 items, 7 each related to the symptoms of anxiety (HAD-A) and depression (HAD-D). Cronbach's alpha coefficient of the scale is 0.77 and the test–retest reliability is 0.73. The higher the score, the greater are anxiety symptoms. A score <8 points indicates lack of any mental disorder, 8–10 points to the possibility of a disorder, and >10 indicates that a disorder is highly likely to be present (Bjelland et al. 2002; Zigmond and Snaith 1983). The scale does not evaluate somatic symptoms, but it detects item bias and provides valid comparisons of patients' well-being in clinical practice (Verdam et al. 2017).

The Zubrod score was used to assess the overall health and performance of NSCLC patients. The score runs from 0 to 5, with 0 denoting perfect health and 5 death (Oken et al. 1982).

2.2 Statistical Elaboration

Continuous data were expressed as means ±SD and categorical data as counts and percentages, and medians with minimum-maximum and interquartile ranges. Data distribution was checked with the Shapiro-Wilk test. Differences between parameters and variables were assessed with Student's *t*-test, Mann-Whitney's U test, Fisher's exact test, or Chi-squared test as required. A p-value <0.05 defined statistically significant changes.

3 Results

3.1 Patients' Characteristics

There were significant differences among the three groups of patients, based on the strategy of coping with stress, regarding the age, forced vital capacity of lungs (FVC), relationship status, performance status, receiving radiation therapy or symptomatic treatment, chest pain, and nutritional status. Patients with constructive coping strategies were older than those with balanced and destructive strategies (66 vs. 63 vs. 61 years, respectively) and had a lower FVC (3.2 vs. 2.9 vs. 2.9 L, respectively; $p < 0.05$). Coping strategies differed alongside the patients' relationship status. Patients with constructive strategies also were more often in a relationship when compared with the other groups (75.0% vs. 50.0% vs. 36.6%, respectively), while patients who were single most often belonged to those using destructive coping strategies (25.0% vs. 50.0% vs. 63.4%, respectively). Similar relationships were found for performance status, as patients with poorer performance status were more likely to have mostly destructive coping strategy, while those with better performance status had mostly constructive coping strategies. Patients also had different coping strategies depending on the treatment received. Those receiving radiation therapy or symptomatic treatment used mostly destructive coping strategies, while those treated surgically had mostly constructive ones. Another significant factor was nutrition. Patients with a normal body weight were more likely to use constructive coping strategies, malnourished patients tended to use destructive strategies, and those at risk of malnutrition had a balanced coping strategy (Table 1).

3.2 Study Outcomes

3.2.1 Anxiety and Depressive Symptoms

In the HADS questionnaire, anxiety and depressive symptoms were associated with negative strategies of coping with cancer. Anxiety scored 10.9 for destructive versus 9.3 for balanced versus 6.3 points for constructive strategy. A similar relationship was found for depressive symptoms, with the highest score in patients with mostly destructive strategies, and the lowest in those with constructive strategy; 11.9 for destructive versus 8.8 for balanced versus 5.8 points for constructive strategy (Table 2).

The intensity of both anxiety and depressive symptoms in NSCLC patients also significantly correlated with the results of lung function tests, pain intensity, assessed on VAS, and the overall number of symptoms. These correlations were positive, but rather weak (Table 3).

3.2.2 Influence of Selected Categorical Variables on Anxiety and Depressive Symptoms

Univariate Analysis The intensity of anxiety and depressive symptoms was affected by some of the nominal and ordinal variables investigated in NSCLC patients. Anxiety was intensified in patients with high school education (compared with vocational education), poor performance status (Zubrod score 2–4), cancer stage T2 and T4 (compared to stage T1), comorbidities, lack of surgical treatment, presence of radiation or symptomatic therapy, and with malnutrition (see

Table 1 Socio-clinical characteristics of NSCLC patients, broken down by strategy for coping with cancer

	Coping strategy			
	Constructive	Balanced	Destructive	*p*
Parameters	mean ± SD			
Age (years)	61 ± 11	63 ± 8	66 ± 8	0.014 MW
FEV1 (L)	2.5 ± 0.8	2.3 ± 0.7	2.2 ± 0.9	0.078 MW
FVC (L)	3.2 ± 1.0	2.9 ± 0.9	2.9 ± 0.9	0.034 MW
FEV1/%FVC	67.5 ± 17.5	69.9 ± 12.0	69.7 ± 7.8	0.897 MW
VAS	3.8 ± 2.1	4.3 ± 1.8	4.7 ± 1.7	0.084 MW
	***n*(%)**			
Female	39 (44.3)	28 (50.0)	17 (41.5)	0.678 Chi^2
Male	49 (55.7)	28 (50.0)	24 (58.5)	
In a relationship	66 (75.0)	28 (50.0)	15 (36.6)	<0.001 Chi^2
Singles	22 (25.0)	28 (50.0)	26 (63.4)	
Primary education	6 (6.8)	2 (3.6)	7 (17.1)	0.151 F
Vocational	39 (44.3)	32 (57.1)	19 (46.3)	
High school	32 (36.4)	19 (33.9)	10 (24.4)	
College/University	11 (12.5)	3 (5.4)	5 (12.2)	
Zubrod score 0	24 (27.3)	7 (12.5)	3 (7.3)	<0.001 F
1	44 (50.0)	25 (44.6)	11 (26.8)	
2	19 (21.6)	21 (37.5)	21 (51.2)	
3	1 (1.1)	3 (5.4)	5 (12.2)	
4	0 (0.0)	0 (0.0)	1 (2.4)	
T1	10 (11.4)	5 (8.9)	5 (12.2)	0.497 F
T1a	8 (9.1)	2 (3.6)	0 (0.0)	
T1b	7 (8.0)	2 (3.6)	2 (4.9)	
T2	19 (21.6)	20 (35.7)	8 (19.5)	
T2a	8 (9.1)	5 (8.9)	4 (9.8)	
T2b	3 (3.4)	4 (7.1)	5 (12.0)	
T3	10 (11.4)	7 (12.5)	4 (9.7)	
T4	22 (25.0)	10 (17.9)	12 (29.3)	
Tx	1 (1.1)	1 (1.8)	0 (0.0)	
N0	35 (39.8)	19 (33.9)	14 (34.2)	0.959 F
N1	19 (21.6)	13 (23.2)	10 (24.4)	
N2	24 (27.3)	17 (30.4)	11 (26.8)	
N3	6 (6.8)	2 (3.6)	3 (7.3)	
Nx	4 (4.6)	5 (8.9)	2 (4.9)	
M0	59 (67.1)	40 (71.4)	29 (70.7)	0.704 F
M1	11 (12.5)	5 (8.9)	5 (12.2)	
M1a	1 (1.1)	1 (1.8)	0 (0.0)	
M1b	7 (8.0)	1 (1.8)	2 (4.9)	
M1x	0 (0.0)	1 (1.8)	0 (0.0)	
M2	1 (1.1)	2 (3.6)	1 (2.4)	
M3	0 (0.0)	0 (0.0)	1 (2.4)	
Mx	9 (10.2)	6 (10.7)	2 (4.9)	
Surgical treatment	63 (71.6)	43 (76.8)	26 (63.4)	0.354 Chi^2
Radiation therapy	19 (21.6)	23 (41.1)	13 (31.7)	0.043 Chi^2
Chemotherapy	46 (52.3)	31 (55.4)	25 (61.0)	0.651 Chi^2
Symptomatic treatment	3 (3.4)	9 (16.1)	4 (9.8)	0.029 F
Unconventional treatment	0 (0.0)	2 (3.6)	1 (2.4)	0.142 F

(continued)

Table 1 (continued)

Parameters	Coping strategy			
	Constructive	Balanced	Destructive	*p*
	mean ± SD			
	***n*(%)**			
Chronic cough	69 (78.4)	46 (82.1)	34 (82.9)	0.78 Chi^2
Shortness of breath	48 (54.6)	37 (66.1)	29 (70.7)	0.152 Chi^2
Chest pain	23 (26.1)	28 (50.0)	23 (56.1)	0.001 Chi^2
Hemoptysis	25 (28.4)	20 (35.7)	12 (29.3)	0.633 Chi^2
Malnutrition	38 (43.2)	26 (46.4)	31 (75.6)	0.003 Chi^2
Risk of malnutrition	20 (22.7)	18 (32.1)	6 (14.6)	
Normal nutrition	30 (34.1)	12 (21.4)	4 (9.8)	

FVC forced vital capacity, *FEV1* forced expiratory volume in 1 s, *VAS* visual-analog scale, *TNM* staging system for cancer where T stands for tumor size, N stands for metastases to lymph nodes, and M stands for metastases to other parts of the body, Chi^2 chi-squared test, *F* Fisher's exact test, *MW* Mann-Whitney U test

Table 2 Anxiety and depressive symptoms in NSCLC patients, broken down by strategy for coping with cancer

HADS	Coping strategy	n	Median	Min-Max	Q3–Q1	p
Anxiety	Destructive (D)	41	12	0–19	14–9	< 0.001 D >B > C
	Balanced (B)	56	10	0–15	12.3–6.8	
	Constructive (C)	88	6.5	0–18	10–2	
Depression	Destructive (D)	41	12	3–18	13–10	< 0.001 D > B > C
	Balanced (B)	56	10.5	0–16	12–7	
	Constructive (C)	88	5.5	0–15	10–0.8	

HADS Hospital Anxiety and Depression Scale (Mann-Whitney test)

Table 3 Influence of lung function and pain intensity in NSCLC patients on the level of anxiety

Variable	Anxiety		Depression	
	r	*p*	*r*	*p*
FEV1	−0.009	0.902	−0.024	0.750
FVC	0.017	0.823	−0.05	0.503
FEV1/%FVC	−0.103	0.162	−0.128	0.083
VAS	0.492	<0.001	0.432	<0.001
Number of symptoms	0.250	0.001	0.247	0.001

FVC forced vital capacity, *FEV1* forced expiratory volume in 1 s, *VAS* visual-analog scale, *r* Spearman's rank correlation coefficient

Table 4 for details). Likewise, depressive symptoms were intensified in patients with poor performance status (Zubrod score 2–4), comorbidities, lack of surgical treatment, presence of radiation and chemotherapy, and with malnutrition or the risk for it (see details in Table 5).

Multivariate Regression Analysis Pain intensity, measured on VAS, was an independent predictor of anxiety: one point increase on the pain scale increased anxiety by a mean of 0.51 points. Further, having at least one comorbidity increased the anxiety score by a mean of 1.29 points. Symptomatic treatment decreased anxiety score by a mean of 2.60 points. Normal nutritional status decreased the anxiety score by a mean of 4.34 points when compared to malnutrition. Additionally, anxiety was affected by the patient's coping strategy: constructive strategies decreased anxiety score by a mean of 2.83 points when compared to destructive strategies (see details in Table 6).

Table 4 Effects of categorical variables on the level of anxiety in NSCLC patients

Group	*n*	Median	Min–Max	Q3–Q1	*p*
Female	84	9	0–19	12–6	0.540
Male	101	9	0–15	12–4	
Aged< 60	71	10	0–18	12.5–6	0.057
Aged >60	114	9	0–19	11–4	
In a relationship	109	9	0–19	12–4	0.803
Singles	76	9	0–15	11.3–6	
Primary education	15	9	0–14	11.5–6	0.030 H > V
Vocational (V)	90	8	0–15	11–3	
High school (H)	61	10	0–19	13–7	
College/University	19	10	0–15	11.5–4	
Zubrod score 0	34	8	0–18	9.8–3	0.002 2–4 > 0
1	80	9	0–15	11–3	
2–4	71	10	0–19	13–7	
T1	41	6	0–18	10–2	0.037 T2, T3, & T4 > T1
T2	76	9	0–15	13–4.8	
T3	21	9	0–15	12–3	
T4	44	10	0–19	12–7	
N0	68	8	0–15	11–2.8	0.057
N1	42	10	0–15	13–6.3	
N2–3	63	10	0–19	12–6	
Nx	11	10	0–15	12–6.5	
M0	128	9	0–19	12–4	0.877
M1–3	38	10	0–18	11.8–4.5	
Mx	18	10	0–15	11–8.3	
No comorbidity	62	4.5	0–19	9.8–1	< 0.001 1 & 2–3 > No
1 disease	82	10	0–15	12–7	
2–3 diseases	41	10	0–15	12–8	
No surgical treatment	53	10	0–15	12–8	0.007
Surgical treatment	132	9	0–19	12–3	
No radiation therapy	130	8	0–19	11–3	< 0.001
Radiation therapy	55	11	0–15	13–9	
No chemotherapy	83	6	0–19	9.5–0	< 0.001
Chemotherapy	102	10	0–15	12–8.3	
No symptomatic treatment	169	9	0–19	12–4	0.107
Symptomatic treatment	16	10.5	6–15	12–8.8	
Malnutrition (M)	95	10	3–19	13–9	< 0.001 M & R > N
Risk of malnutrition (R)	44	9.5	0–15	12–4	
Normal nutrition (N)	46	3	0–15	6–0	

TNM staging system for cancer where T stands for tumor size, N stands for metastases to lymph nodes, and M stands for metastases to other body parts (Kruskal-Wallis followed by *post-hoc* Dunn test)

Likewise, pain intensity also was an independent predictor of the intensity of depressive symptoms in multivariate analysis; one point increase on the pain scale increased the depressive symptoms score by a mean of 0.41 points. College/university education increased the score by a mean of 3.05 points, when compared to primary education. Another determinant was performance status, with poorer performance associated with a higher score of depressive symptoms; Zubrod scores of 2–4 increased the score by a mean of 2.16 points. Chemotherapy increased the score by a mean of

Table 5 Effects of categorical variables on the level of depressive symptoms in NSCLC patients

Group	*n*	Median	Min–Max	Q3–Q1	*p*
Female	84	9	0–18	12–4	0.985
Male	101	9	0–18	12–3	
Aged < 60	71	10	0–15	12–4.5	0.552
Aged > 60	114	9	0–18	12–3	
In a relationship	109	9	0–18	12–2	0.198
Singles	76	10	0–18	12–6	
Primary education	15	8	0–18	10.5–4	0.144
Vocational	90	8	0–15	12–1.3	
High school	61	10	0–18	13–7	
College/University	19	11	0–17	12.5–6	
Zubrod score 0	34	6	0–15	9–3	< 0.001 2–4 > 0
1	80	8.5	0–15	12–1	
2–4	71	11	0–18	13–8	
T1	41	7	0–15	11–1	0.061
T2	76	9	0–18	12–3	
T3	21	9	0–18	12–1	
T4	44	11	0–18	12.3–7.8	
N0	68	8	0–15	11–1	0.084
N1	42	9.5	0–18	13–6.3	
N2–3	63	10	0–18	12.5–5	
Nx	11	10	0–13	11.5–8	
M0	128	9	0–18	12–3	0.501
M1–3	38	10.5	0–17	12–4.5	
Mx	18	9.5	0–15	13–6.3	
No comorbidity	62	5	0–15	9.8–0	< 0.001 1 & 2–3 > No
1 disease	82	10	0–18	12–6	
2–3 diseases	41	11	0–18	13–8	
No metastases	121	9	0–15	11–3	0.073
Metastases	64	11	0–18	13–4	
No surgical treatment	53	11	0–18	12–8	0.006
Surgical treatment	132	8	0–18	12–1	
No radiation therapy	130	8	0–18	11.8–1	0.001
Radiation therapy	55	11	0–15	13–8	
No chemotherapy	83	5	0–18	10–0	< 0.001
Chemotherapy	102	11	0–18	13–8	
No symptomatic treatment	169	9	0–18	12–3	0.051
Symptomatic treatment	16	11	4–16	12.3–8.8	
Malnutrition (M)	95	11	0–18	13–8	< 0.001 M & R > N
Risk of malnutrition (R)	44	9.5	0–17	12–3.8	
Normal nutrition (N)	46	0	0–15	7.8–0	

TNM staging system for cancer where T stands for tumor size, N stands for metastases to lymph nodes, and M stands for metastases to other body parts (Kruskal-Wallis followed by *post-hoc* Dunn test)

2.36 points. Normal nutrition decreased the score by a mean of 3.21 points, when compared to malnutrition. Coping strategy also was an independent determinant of depressive symptoms. Constructive coping strategy with cancer decreased the score by a mean of 3.87 points and balanced strategy 1.88 points, when compared to destructive strategy (see details in Table 7).

Table 6 Effects of categorical variables on the level of anxiety in NSCLC patients in multivariate regression model

	Regression coefficient	95% CI		*p*
FEV1	−0.356	−1.409	0.698	0.505
FVC	0.482	−0.356	1.319	0.257
FEV1/%FVC	0.021	−0.022	0.064	0.337
VAS	0.512	0.235	0.789	<0.001
Number of symptoms	−0.064	−0.476	0.348	0.759
Female	Reference item			
Male	−0.462	−1.562	0.638	0.408
Aged <60	Reference item			
Aged >60	−1.048	−2.191	0.096	0.072
In a relationship	Reference item			
Singles	−0.313	−1.476	0.851	0.596
Primary Education	Reference item			
Vocational	−0.500	−2.443	1.443	0.612
High school	1.210	−0.881	3.302	0.255
College/university	0.680	−1.925	3.285	0.607
Zubrod score 0	Reference item			
1	0.569	−0.892	2.030	0.443
2–4	1.534	−0.064	3.132	0.060
T1	Reference item			
T2	0.599	−0.757	1.955	0.384
T3	0.502	−1.529	2.532	0.626
T4	−0.012	−1.906	1.881	0.990
N0	Reference item			
N1	−0.188	−1.639	1.263	0.798
N2–3	0.003	−1.432	1.438	0.997
Nx	0.117	−2.497	2.731	0.930
M0	Reference item			
M1–3	−0.535	−2.051	0.980	0.486
Mx	−0.148	−2.063	1.766	0.879
No comorbidities	Reference item			
1 Disease	1.293	0.005	2.580	0.049
2–3 Diseases	0.622	−0.996	2.240	0.449
No metastases	Reference item			
Metastases	−0.100	−1.477	1.276	0.886
No surgical treatment	Reference item			
Surgical treatment	−0.824	−2.535	0.887	0.343
No radiation therapy	Reference item			
Radiation therapy	0.953	−0.386	2.292	0.162
No chemotherapy	Reference item			
Chemotherapy	1.273	−0.195	2.740	0.089
No symptomatic treatment	Reference item			
Symptomatic treatment	−2.604	−4.622	−0.587	0.012
Malnutrition	Reference item			
Risk of malnutrition	−1.244	−2.635	0.147	0.079
Normal nutrition	−4.343	−5.77	−2.917	<0.001
Destructive coping strategy	Reference item			
Balanced coping strategy	−0.709	−2.145	0.726	0.330
Constructive coping strategy	−2.830	−4.255	−1.405	<0.001

FVC forced vital capacity, *FEV1* forced expiratory volume in 1 s, *VAS* visual-analog scale, *TNM* staging system for cancer where T stands for tumor size, N stands for metastases to lymph nodes, and M stands for metastases to other parts of the body. R^2 coefficient for the model was 63.7%, meaning that variables included in the model accounted for 63.7% of variance in anxiety scores. The remaining 36.3% depended on variables not included in the model or on random factors

Table 7 Effects of categorical variables on the level of anxiety in NSCLC patients in multivariate regression model

	Regression coefficient	95% CI		*p*
FEV1	−0.034	−1.231	1.162	0.955
FVC	−0.172	−1.123	0.779	0.722
FEV1/%FVC	0.013	−0.036	0.062	0.597
VAS	0.407	0.093	0.722	0.011
Number of symptoms	−0.113	−0.582	0.355	0.633
Female	Reference item			
Male	0.372	−0.877	1.622	0.557
Aged <60	Reference item			
Aged >60	−0.435	−1.734	0.863	0.509
In a relationship	Reference item			
Singles	0.935	−0.387	2.256	0.164
Primary education	Reference item			
Vocational	0.365	−1.842	2.572	0.744
High school	1.888	−0.487	4.263	0.118
College/university	3.045	0.087	6.004	0.044
Zubrod score 0	Reference item			
1	0.711	−0.948	2.370	0.399
2–4	2.163	0.348	3.978	0.020
T1	Reference item			
T2	0.681	−0.859	2.220	0.384
T3	0.990	−1.317	3.296	0.398
T4	0.377	−1.773	2.528	0.729
N0	Reference item			
N1	0.451	−1.197	2.099	0.589
N2–3	0.214	−1.416	1.844	0.796
Nx	−1.234	−4.203	1.734	0.413
M0	Reference item			
M1–3	0.322	−1.400	2.043	0.712
Mx	−0.036	−2.210	2.138	0.974
No comorbidities	Reference item			
1 Disease	1.033	−0.430	2.495	0.165
2–3 Diseases	1.813	−0.025	3.651	0.053
No metastases	Reference item			
Metastases	−0.524	−2.088	1.039	0.509
No surgical treatment	Reference item			
Surgical treatment	−0.294	−2.237	1.649	0.766
No radiation therapy	Reference item			
Radiation therapy	−0.154	−1.675	1.367	0.842
No chemotherapy	Reference item			
Chemotherapy	2.362	0.696	4.029	0.006
No symptomatic treatment	Reference item			
Symptomatic treatment	−1.951	−4.242	0.341	0.095
Malnutrition	Reference item			
Risk of malnutrition	−0.014	−1.593	1.566	0.986
Normal nutrition	−3.206	−4.826	−1.585	<0.001
Destructive coping strategy	Reference item			
Balanced coping strategy	−1.882	−3.512	−0.252	0.024
Constructive coping strategy	−3.869	−5.488	−2.251	<0.001

FVC forced vital capacity, *FEV1* forced expiratory volume in 1 s, *VAS* visual-analog scale, *TNM* staging system for cancer where T stands for tumor size, N stands for metastases to lymph nodes, and M stands for metastases to other parts of the body. The R^2 coefficient for the model was 58.6%, meaning that variables included in the model accounted for 58.6% of variance in the score of depressive symptoms. The remaining 41.4% depended on variables not included in the model or on random factors

4 Discussion

The present study demonstrates that dominant emotions experienced by patients who are faced with cancer included anxiety and depression, and patients exhibit one of two opposite attitudes toward the new situation. Constructive attitude aims at fighting for life and health, and involves adaptive behaviors. Contrarily, destructive attitude typically results in resignation, maladaptive behaviors, and passive submission to the disease. In the latter case, patients show anxious preoccupation, and a sense of helplessness and hopelessness. The manner in which patients cope with a stressful situation substantially factors in quality of life (Chabowski et al. 2018a). Destructive coping strategies have an impact on quality of life and treatment outcomes, while constructive strategies are associated with lower pain and stress levels and better overall health. Literature on the subject matter demonstrates that most lung cancer patients exhibit emotion-focused strategies, which are less effective than problem-focused strategies (Genç and Tan 2011; Horney et al. 2011; Kershaw et al. 2004).

The present findings confirmed the assumption that mental adjustment and strategies for coping with cancer would be significantly associated with the level of negative emotions experienced. Patients using constructive strategies had less intense symptoms of depressive symptoms and anxiety than those using destructive strategies. In multivariate regression analysis, constructive coping strategy was found beneficial with regard to lowering depressive and anxiety symptoms scores; by a mean of 3.87 and 2.83 points, respectively. These scores evidently increased with negative attitudes. Shimizu et al. (2015) have shown that psychological factors have so strong a bearing on the level of anxiety that none of the cancer-related variables included into multivariate analysis such as personal characteristics, health behaviors, or physical symptoms remain of meaningful significance. In the present study, aside from psychological aspects of coping, significant factors also included socio-clinical variables, which strongly influenced negative emotions experienced by patients. Anxiety was affected by pain intensity, comorbidities, symptomatic treatment, and nutritional status. Depression, on the other side, was affected by college/university education, performance status, chemotherapy, and nutritional status. A study by Polański et al. (2018) has shown an association between anxiety and depression, on the one hand, and lower quality of life and more severe symptoms in lung cancer patients, on the other hand. Those authors also demonstrate that patients with problem-focused coping strategies experience less pain, shortness of breath, vomiting, insomnia, coughing, and diarrhea. Likewise, Salvo et al. (2012) have found that disease symptoms associate with anxiety. Faller and Schmidt (2004) have listed cancer stage, cancer type, gender, and age among the factors with the strongest influence on depressive symptoms. Others have also shown that NSCLC patients focusing on emotions experience considerably more psychological distress and more severe clinical symptoms, for instance, nausea (Prasertsri et al. 2011; Kuo and Ma 2002) as opposed to problem-focused coping strategies that help decrease the intensity of symptoms, particularly depression, although less so concerning anxiety (Shimizu et al. 2015). We found no such dichotomy in the present study as constructive strategy of coping with cancer alleviated both depression and anxiety.

Studies suggest that women are easily affected by anxiety and depression, although the underlying mechanisms of gender differences are unclear (Wang et al. 2017). In the present study, female patients did score more on anxiety and depression than male patients, but the difference failed to reach statistical significance. Likewise, no gender differences were noticed concerning the coping strategies, nor was gender a predictor of negative emotions in multivariate analysis.

Nutritional status was identified as a determinant of negative emotions in the present study. Patients with destructive coping strategy tended to be malnourished, while those with a predominantly constructive strategy, on average, had normal nutritional status. In univariate analysis, malnutrition was correlated with anxiety and depression levels, and in multivariate analysis, normal nutritional status was a significant determinant of lower anxiety and depression scores. Chabowski et al. (2018b) have demonstrated that normal nutritional status correlates with less perceived pain and lower anxiety and depression in NSCLC patients when compared to the presence of nutritional disorders. This correlation implies that poor nutrition is among factors that may lead to increased morbidity and mortality of lung cancer patients. In line with those results, Polański et al. (2017) have shown that malnutrition is an independent determinant of decreased quality of life in the domain of physical functioning.

Studies on the coping strategies in lung cancer patients are rather scarce. Chabowski et al. (2018a) have demonstrated an association between constructive coping strategies and improved quality of life. Mosher et al. (2015) have provided preliminary data on strategies used by patients with advanced, symptomatic lung cancer and their primary family caregivers to cope with physical and psychological cancer symptoms. A number of patients reported a way of coping by maintaining a "normal routine". Cognitive methods of coping were also reported, such as managing expectations, fostering a positive attitude, and avoiding thinking about the illness. Based on these reports, patients who experience cancer symptoms seem to avoid negative emotions rather than to process and express them.

In the present study, type of treatment used was another predictor of anxiety and depression levels, identified in multivariate analysis. Patients treated surgically were more likely to have a constructive coping strategy than those treated with chemotherapy or radiation therapy. Symptomatic treatment significantly reduced the patients' anxiety, while chemotherapy adversely correlated with depressive symptoms. Chemotherapy and radiation therapy apparently increase the sense of uncertainty and danger in patients, resulting in emotional disorders that may lead to depression and anxiety.

Currently, surgery is the treatment of choice and the most effective method, with twice the success rate when compared with chemotherapy or radiation therapy. Chemotherapy, which is rather commonly administered, is associated with more intense pain, respiratory complaints, and fatigue. Lung cancer patients treated with chemotherapy have a lower quality of life, although up to 50% of patients receive the treatment, according to studies. Each consecutive run of chemotherapy is associated with significantly reduced quality of life (Wintner et al. 2013). Milošević et al. (2016) also have reported a significant deterioration in functioning among patients receiving symptomatic treatment. Noticeably, however, quality of life deterioration may result from disease progression rather than from specific treatment.

Performance status, assessed with the Zubrod score, was another significant determinant of anxiety and depressive symptoms in the present study. In univariate analysis, higher Zubrod scores were associated with higher levels of both anxiety and depressive symptoms, while in multivariate analysis, a higher Zubrod score was a significant predictor of depressive symptoms only. In the literature, a similar negative influence has been found for fatigue which can result from both the underlying disease and treatment used. Fatigue, a symptom that may affect up to 80% of patients, is often disregarded by medical personnel, although it can significantly reduce functional performance, especially in conjunction with other symptoms such as pain and sleep disorders. Hopwood and Stephens (2000) have confirmed the presence of an association between functional deficits and the development of depressive symptoms.

5 Summary and Conclusion

The levels of anxiety and depressive symptoms are lower in non-small cell lung cancer patients who exhibit constructive strategies for coping with a chronic illness. A constructive coping strategy is a significant independent determinant of anxiety and depressive symptoms in patients with lung cancer. Other predictors of lower anxiety and depression include symptomatic treatment and a good nutritional status, while pain, chemotherapy, and poor performance status exacerbate the negative emotions.

We conclude that early identification of anxiety and depression has an important bearing on the choice of treatment and the treatment process itself. Thus, psychological assessment is essential in clinical care for patients with lung cancer. Additionally, the way the patient adjusts mentally to having a cancer disease, expressed in the coping strategy, helps recognize and fulfill his/her individual needs. Both screening tests and regular follow-ups in patients diagnosed with lung cancer should be performed to identify those with maladaptive coping strategies and to provide them with psychotherapeutic assistance. Notably, reverting to a constructive coping strategy may not only reduce negative emotions, but also alleviate symptoms and improve outcome of cancer illness.

Conflicts of Interest The authors declare no conflicts of interest in relation to this article.

Ethical Approval All procedures performed in studies involving human participants were in accordance with the ethical standards of the institutional and/or national research committee and with the 1964 Helsinki declaration and its later amendments or comparable ethical standards. The study was approved by the Bioethics Committee of Wroclaw Medical University in Poland (permit no. 507/2015).

Informed Consent Informed consent was obtained from all individual participants included in the study.

References

Arrieta O, Angulo LP, Núñez-Valencia C, Dorantes-Gallareta Y, Macedo EO, Martinez-Lopez D, Alvarado S, Corona-Cruz JF, Onate-Ocana LF (2013) Association of depression and anxiety on quality of life, treatment adherence, and prognosis in patients with advanced non-small cell lung cancer. Ann Surg Oncol 20(6):1941–1948

Bjelland I, Dahl AA, Haug TT, Neckelmann D (2002) The validity of the hospital anxiety and depression scale. An updated literature review. J Psychosom Res 52 (2):69–77

Bray F, Ferlay J, Soerjomataram I, Siegel RL, Torre LA, Jemal A (2018) Global cancer statistics 2018: GLOBOCAN estimates of incidence and mortality worldwide for 36 cancers in 185 countries. CA Cancer J Clin 68(6):394–424

Brown LF, Kroenke K, Theobald DE, Wu J, Tu W (2010) The association of depression and anxiety with health-related quality of life in cancer patients with depression and/or pain. Psychooncology 19:734–741

Carlsen K, Jensen AB, Jacobsen E, Krasnik M, Johansen C (2005) Psychosocial aspects of lung cancer. Lung Cancer 47(3):293–300

Carlson LE, Zelinski EL, Toivonen KI, Sundstrom L, Jobin CT, Damaskos P, Zebrack B (2019) Prevalence of psychosocial distress in cancer patients across 55 North American cancer centers. J Psychosoc Oncol 37(1):5–21

Chabowski M, Jankowska-Polańska B, Lomper K, Janczak D (2018a) The effect of coping strategy on quality of life in patients with NSCLC. Cancer Manag Res 10:4085–4093

Chabowski M, Polański J, Jankowska-Polańska B, Janczak D, Rosińczuk J (2018b) Is nutritional status associated with the level of anxiety, depression and pain in patients with lung cancer? J Thorac Dis 10 (4):2303–2310

Czerw AI, Marek E, Deptała A (2015) Use of the mini-MAC scale in the evaluation of mental adjustment to cancer. Contemp Oncol (Pozn) 19(5):414–419

Faller H, Schmidt M (2004) Prognostic value of depressive coping and depression in survival of lung cancer patients. Psychooncology 13:359–363

Fitzell A1, Pakenham KI (2010) Application of a stress and coping model to positive and negative adjustment outcomes in colorectal cancer caregiving. Psychooncology 19(11):1171–1178

Genç F, Tan M (2011) Symptoms of patients with lung cancer undergoing chemotherapy and coping strategies. Cancer Nurs 34(6):503–509

Gonzalez BD, Jacobsen PB (2012) Depression in lung cancer patients: the role of perceived stigma. Psychooncology 21(3):239–246

Greer JA, Pirl WF, Park ER, Lynch TJ, Temel JS (2008) Behavioral and psychological predictors of chemotherapy adherence in patients with advanced non-small cell lung cancer. J Psychosom Res 65:549–552

Grønberg BH, Sundstrøm S, Kaasa S, Bremnes RM, Fløtten O, Amundsen T, Hjelde HH, Plessen CV (2010) Influence of comorbidity on survival, toxicity and health-related quality of life in patients with advanced non-small-cell lung cancer receiving platinum-doublet chemotherapy. Eur J Cancer 46 (12):2225–2234

Hopwood P, Stephens RJ (2000) Depression in patients with lung cancer: prevalence and risk factors derived from quality-of-life data. J Clin Oncol 18(4):893–903

Horney DJ, Smith HE, McGurk M, Weinman J, Herold J, Altman K, Llewellyn CD (2011) Associations between quality of life, coping strategies, optimism, and anxiety and depression in pretreatment patients with head and neck cancer. Head Neck 33(1):65–71

House A, Stark D (2002) Anxiety in medical patients. BMJ 325:207–209

Iyer S, Taylor-Stokes G, Roughley A (2013) Symptom burden and quality of life in advanced non-small cell lung cancer patients in France and Germany. Lung Cancer 81(2):288–293

Juczyński Z (2001) Measurement tools in health promotion and psychology, 2nd edn. Laboratory of Psychological Tests of the Polish Psychological Association, Warsaw

Kershaw T, Northouse L, Kritpracha C, Schafenacker A, Mood D (2004) Coping strategies and quality of life in women with advanced breast cancer and their family caregivers. Psychol Health 19(2):139–155

Kim Y, Duberstein PR, Sörensen S, Larson MR (2005) Levels of depressive symptoms in spouses of people with lung cancer: effects of personality, social support, and caregiving burden. Psychosomatics 46:123–130

Kuo TT, Ma FC (2002) Symptom distresses and coping strategies in patients with non-small cell lung cancer. Cancer Nurs 25(4):309–317

Latini DM, Hart SL, Knight SJ, Cowan JE, Ross PL, Duchane J, Carroll PR, CaPSURE Investigators (2007) The relationship between anxiety and time to treatment for patients with prostate cancer on surveillance. J Urol 178(3Pt 1):826–831

Lazarus RS, Folkman S (1984) In stress, appraisal, and coping. Springer, New York

Milošević B, Pejić D, Momčičević D, Kovačević P, Stanetić M, Dragić S (2016) Quality of life in lung cancer patients due to treatment. Signa Vitae 11(Suppl 2):47–50

Montazeri A, Milroy R, Hole D, McEwen J, Gillis CR (2001) Quality of life in lung cancer patients: as an important prognostic factor. Lung Cancer 31 (2–3):233–240

Mosher CE, Ott MA, Hanna N, Jalal SI, Champion VL (2015) Coping with physical and psychological symptoms: a qualitative study of advanced lung cancer patients and their family caregivers. Support Care Cancer 23(7):2053–2060

Oken MM, Creech RH, Tormey DC, Horton J, Davis TE, McFadden ET, Carbone PP (1982) Toxicity and response criteria of the Eastern Cooperative Oncology Group. Am J Clin Oncol 5(6):649–655

Parkin DM, Pisani P, Ferlay J (1999) Global cancer statistics. CA Cancer J Clin 49(1):33–64

Pirl WF, Greer JA, Traeger L, Jackson V, Lennes IT, Gallagher ER, Perez-Cruz P, Heist RS, Temel JS (2012) Depression and survival in metastatic non-small-cell lung cancer: effects of early palliative care. J Clin Oncol 30(12):1310–1315

Polański J, Jankowska-Polańska B, Uchmanowicz I, Chabowski M, Janczak D, Mazur G, Rosińczuk J (2017) Malnutrition and quality of life in patients with non-small-cell lung cancer. Adv Exp Med Biol 1021:15–26

Polański J, Chabowski M, Chudiak A, Uchmanowicz B, Janczak D, Rosińczuk J, Mazur G (2018) Intensity of anxiety and depression in patients with lung cancer in relation to quality of life. Adv Exp Med Biol 1023:29–36

Prasertsri N, Holden J, Keefe FJ, Wilkie DJ (2011) Repressive coping strategy: relationships with depression, pain, and pain coping strategies in lung lancer outpatient. Lung Cancer 71(2):235–240

Sak J, Sagan D, Wiechetek M, Pawlikowski J (2012) Suboptimal perception of illness due to self-realization constraints impairs psychological welfare in surgical patients. Eur J Cardiothorac Surg 41(4):824–828

Salvo N, Zeng L, Zhang L, Leung M, Khan L, Presutti R, Nguyen J, Holden L, Culleton S, Chow E (2012) Frequency of reporting and predictive factors for anxiety and depression in patients with advanced cancer. Clin Oncol (R Coll Radiol) 24(2):139–148

Shimizu K, Nakaya N, Saito-Nakaya K, Akechi T, Ogawa A, Fujisawa D, Sone T, Yoshiuchi K, Goto K, Iwasaki M, Tsugane S, Uchitomi Y (2015) Personality traits and coping strategies explain anxiety in lung cancer patients to a greater extent than other factors. Jpn J Clin Oncol 45(5):456–463

Vaillo AY, Pérez MS, López MP, Retes RR (2018) Mini-mental adjustment to cancer scale: construct validation in Spanish breast cancer patients. J Psychosom Res 114:38–44

Verdam MGE, Oort FJ, Sprangers MAG (2017) Item bias detection in the Hospital Anxiety and Depression Scale using structural equation modeling: comparison with other item bias detection methods. Qual Life Res 26 (6):1439–1450

Wang S, Tang J, Sun T et al (2017) Survival changes in patients with small cell lung cancer and disparities between different sexes, socioeconomic statuses and ages. Sci Rep 7(1):1339

Watson M, Greer S, Young J, Inayat Q, Burgess C, Robertson B (1988) Development of a questionnaire measure of adjustment to cancer: the MAC scale. Psychol Med 18(1):203–209

Watson M, Law MG, Santos M, Greer S, Baruch J, Bliss J (1994) The Mini-MAC. J Psychosoc Oncol 12 (3):33–46

Wintner LM, Giesinger J, Zabernigg A, Sztankay M, Meraner V, Pall G, Hilbe W, Holzner B (2013) Quality of life during chemotherapy in lung cancer patients: results across different treatment lines. Br J Cancer 109 (9):2301–2308

Zabora J, BrintzenhofeSzoc K, Curbow B, Hooker C, Piantadosi S (2001) The prevalence of psychological distress by cancer site. Psychooncology 10(1):19–28

Zigmond AS, Snaith RP (1983) The hospital anxiety and depression scale. Acta Psychiatr Scand 67(6):361–370

Adv Exp Med Biol - Clinical and Experimental Biomedicine (2020) 8: 71–80
https://doi.org/10.1007/5584_2019_458

Published online: 11 January 2020

Inflammatory Markers During Continuous High Cutoff Hemodialysis in Patients with Septic Shock and Acute Kidney Injury

Grzegorz Kade, Sławomir Literacki, Agnieszka Rzeszotarska, Stanisław Niemczyk, and Arkadiusz Lubas

Abstract

High cut-off (HCO) continuous veno-venous hemodialysis (CVVHD) is one of the renal replacement therapies which nonselectively removes inflammatory mediators. This study seeks to examine the association between the inflammatory background and the need for catecholamine treatment in hemodynamically instable patients having septic shock and acute kidney injury during HCO-CVVHD. There were 38 patients (F/M; 16/22, mean age 63 ± 16 years) included in the study. The initial content of the cytokines IL-4, IL-12, IL-17, and TNFα, C-reactive protein, and the score of the Sequential Organ Failure Assessment (SOFA) were assessed. The receiver operating characteristic (ROC) plot showed that a combination consisting of IL-17 × SOFA ≤22.3 was a reliable predictive factor of the need for catecholamine treatment during HCO-CVVHD, with 82% sensitivity and 90% specificity, with the area under curve (AUC) of 0.843; $p < 0.001$. On the other side, SOFA ≤14.0 predicted catecholamine treatment or its discontinuation when started, with both specificity and sensitivity 83% (AUC = 0.899; $p < 0.001$). In conclusion, the immune system activation, assessed from the initial level of IL-17, and the clinical SOFA evaluation are of practical help in predicting the need for catecholamine treatment or the probability of a reduction thereof in patients on veno-venous hemodialysis due to septic shock.

Keywords

Catecholamine treatment · Cytokines · Hemodialysis · Immune system activation · Kidney injury · Septic shock

G. Kade (✉), S. Niemczyk, and A. Lubas
Department of Internal Diseases, Nephrology and Dialysis, Military Institute of Medicine, Warsaw, Poland
e-mail: gkade5@wp.pl

S. Literacki
Department of Laboratory Diagnostics, Military Institute of Medicine, Warsaw, Poland

A. Rzeszotarska
Department of Clinical Transfusion, Military Institute of Medicine, Warsaw, Poland

1 Introduction

Sepsis remains the main cause of mortality in intensive care units. It also is a major economic burden for health services (Martin–Loeches et al. 2019). High mortality in the course of sepsis is associated with overabundant release of proinflammatory cytokines (Gierek et al. 2011; Harrison et al. 2006). De Vriese et al. (1999)

have hypothesized that removal of the cytokines from circulating blood may reduce organ damage and improve clinical outcome. Therefore, various extracorporeal renal replacement therapies have been introduced in search for the most effective method of removing inflammatory mediators. One such therapy is high cutoff continuous veno-venous hemodialysis (HCO-CVVHD). HCO hemofilters have a larger effective pore size, which is conducive to removal of middle-sized molecules (Morgera et al. 2006).

In a previous study, we have shown that increased content of interleukin-6 (IL-6) during HCO-CVVHD is associated with increased mortality (Kade et al. 2016). IL-6 increases in alarm conditions of the body and induces acute phase reactants, which affects the immune system homeostasis (Xing et al. 1998). IL-6 is a cytokine that may play a dichotomous pro- or anti-inflammatory role. The role in septic shock due to acute kidney injury is uncertain. Therefore, the aim of this study was to get insights into the release of IL-6 and other markers of inflammation during HCO-CVVHD in patients with septic shock and acute kidney injury. We also attempted to investigate the effect of catecholamine treatment on the cytokine content in this condition.

2 Methods

2.1 Patients

The study comprised 38 patients (F/M; 16/22, mean age 63 ± 16 years) with septic shock and acute kidney injury, who were treated with HCO-CVVHD. The diagnosis of septic shock was based on the Surviving Sepsis Campaign 2016 guidelines (Rhodes et al. 2017). The diagnosis of acute kidney injury was based on the Kidney Disease Improving Global Outcomes. Criteria take into account an increase in blood creatinine content and a reduction of diuresis, accompanied by at least one of the following: hypervolemia, increased urea content, hyperkalemia, or a severe disturbance in acid-base balance. The Sequential Organ Failure Assessment (SOFA) was applied in all of the patients. The accumulation of multi-organ dysfunction, including acute kidney injury, was a reason for starting 24-h HCO-CVVHD with a polyarylethersulfone membrane, enabling the removal of cytokines (Septex, Gambro Lundia AB; Lund, Sweden). The membrane effective surface area was 1.1 m^2 and cutoff point 45 kDa. Catecholamines (dopamine, dobutamine, norepinephrine, and epinephrine) were used in patients with persistent hypotension (systolic blood pressure <90 mmHg, mean arterial pressure <65 mmHg, or decrease in systolic blood pressure by >40 mmHg). The catecholamine of choice was norepinephrine.

2.2 Study Protocol

During HCO-CVVHD, the mean blood flow (Q_B) was 149.7 ± 9.7 ml/min. Bicarbonate-buffered fluid was used for all procedures of renal replacement therapy. The mean dialysis fluid flow rate (Q_D), i.e., a volume of dialysis fluid running into the circuit *per* unit of time, was 1197.4 ± 461.8 ml/h. Net ultrafiltration (Q_{UF}^{NET}), i.e., volume removed from the patient by the machine *per* unit of time, was based on the individual fluid status of a patient and, on average, amounted to 56.3 ± 36.1 ml/h. The extracorporeal circuit was anticoagulated by continuous unfractionated heparin infusion according to the individual patient-adjusted regimen. The replacement flow rate in pre-dilution (Q_R^{PRE}) was 250 ml/h. In case of patients with severe bleeding disorders, the dialyzer was flushed with 0.9% NaCl instead of heparin infusion.

Arterial blood pressure was closely monitored in all of the patients. The dose of catecholamine was targeted to the mean arterial pressure of 60 mmHg. When the blood pressure kept on increasing beyond this target, the dose was adjusted by staff of the intensive care unit. Adjustments of catecholamine dose were documented on the hourly basis or more frequently when needed.

The serum content of C-reactive protein (CRP), interleukin-1B (IL-1B), IL-4, IL-6, IL-12, IL-17, interferon gamma (IFNγ), and tumor necrosis factor alpha (TNFα) was measured before and 24 h after therapy. CRP was determined with a nephelometric method (BN II analyzer; Siemens, Munich, Germany), with the norm $\leq$0.5 mg/dl. Cytokine content was determined with a Luminex-Human HS Cytokine Panel A (R&D Systems; Minneapolis, MN). The detection range of individual cytokines was as follows: IL-1B, 0.36–23,150 pg/ml; IL-4, 1.79–29,250 pg/ml; IL-6, 1.66–27,200 pg/ml; IL-12, 1.40–18,650 pg/ml; IL-17, 0.0–1.2 pg/ml; IFNγ, 6.01–6,650 pg/ml; and TNFα, 4.74–14,200 pg/ml.

2.3 Statistical Elaboration

Due to a variable pattern of data distribution, the results were presented as means ±SD or medians with minimum and maximum values. Pearson's and Spearman's correlation coefficients were used for testing correlation of data taken from an interval or ordinal scale, respectively. Accordingly, differences between two groups were tested with a *t*-test or Mann-Whitney U test and between more than two groups with one-way ANOVA or Kruskal-Wallis test. A p-value <0.05 defined statistical significance of differences. Changes in the use of catecholamines also were evaluated. The assumption was made that catecholamine use decreased when there was a reduction in the dose during HCO-CVVHD, it increased in the opposite situation, or it was stable when there were no changes in the dose or the number of dose reductions equaled that of increases. To determine the independent factors connected with the change of catecholamine use during HCO-CVVHD, linear stepwise regression analysis was used. The threshold value of investigated predictors was calculated from the receiver operating characteristic curve (ROC) plots. A commercial statistical package of Statistica v12 software was used (StatSoft Inc., Tulsa, OK).

3 Results

Basic characteristics of the study patients are presented in Table 1. Out of the 38 patients, catecholamines were administered intravenously in 28 patients. Dopamine was infused in 8, dobutamine in 13, norepinephrine in 27, and epinephrine in 11 patients. Four different catecholamines were given in four, three catecholamines in six, and two catecholamines in six patients. Ten patients were hemodynamically stable and did not require any catecholamines (Group 1). The mean, initial, and final doses of each catecholamine administered, with dose changes during HCO-CVVHD, were presented in Table 2.

The doses of catecholamines were decreased in 6 patients (Group 2), remained constant in 17, or increased in 5 patients (Group 3) during HCO-CVVHD. The initial levels of IL-4, IL-12, IFNγ, and SOFA score were significantly lower in Group 2, where there were decreases in catecholamine doses, than those in Group 3, where no decreases in catecholamines were noticed ($p < 0.001$) (Table 3). Decreases in catecholamines in Group 2 correlated with the SOFA score ($r = 0.56$, $p = 0.001$) and with the initial levels of IL-4, IL-12, and IFNγ ($r = 0.46$, $p = 0.011$; $r = 0.53$, $p = 0.003$; and $r = 0.49$, $p = 0.008$; respectively). In the multivariate regression analysis, concerning the patients treated with catecholamines, none of the significantly correlated cytokines, i.e., IL-4, IL-12, or IFNγ, were independently associated with a decrease in catecholamines. However, inclusion of the SOFA score to the regression model

Table 1 Basic characteristics of the study group comprising 16 female and 22 male patients

	Mean ± SD	Median	Minimum	Maximum
Age (years)	63 ± 15.5	63	26	90
Weight (kg)	78.6 ± 20.5	74.5	50	150
Creatinine (mg/dl)	2.8 ± 1.7	2.5	0.6	7.7
SOFA score	14.8 ± 4.2	15.5	6	23

SOFA Sequential Organ Failure Assessment

Table 2 Catecholamine doses administered during high cutoff continuous veno-venous hemodialysis (HCO-CVVHD)

Catecholamine (μg/kg/min)	n	Mean ± SD	Median	Minimum	Maximum
DA–m	8	12.13 ± 11.37	8.61	1.67	32.10
DA (0)	8	8.31 ± 8.36	3.87	1.03	21.67
DA (1)	8	11.01 ± 9.23	9.23	1.67	26.67
Δ DA	8	2.70 ± 3.77	1.43	0.00	10.77
DB–m	12	10.20 ± 9.16	6.42	1.19	31.95
DB(0)	12	10.91 ± 9.31	6.45	1.19	27.68
DB(1)	12	10.24 ± 10.16	5.66	1.19	33.33
Δ DB	12	−0.66 ± 7.95	0.00	−24.43	7.14
NE–m	27	0.72 ± 1.52	0.30	0.05	8.07
NE (0)	27	0.46 ± 0.40	0.30	0.05	1.60
NE (1)	27	0.47 ± 0.45	0.30	0.05	1.87
Δ NE	27	0.02 ± 0.13	0.00	−0.18	0.38
E–m	11	0.21 ± 0.12	0.20	0.02	0.45
E (0)	11	0.20 ± 0.12	0.20	0.01	0.46
E (1)	11	0.22 ± 0.12	0.20	0.01	0.42
Δ E	11	0.02 ± 0.08	0.00	−0.09	0.21

DA dopamine, *DB* dobutamine, *NE* norepinephrine, *E* epinephrine, *m* mean dose during whole HCO-CVVHD, (*0*) initial dose, (*1*) dose at the end of HCO-CVVHD, *Δ* difference between final and initial dose

Table 3 Comparison of the initial immune system activation in groups with different status of catecholamine treatment

	Group 1 (n = 10)	Group 2 (n = 6)	Group 3 (n = 22)
Marker	Mean ± SD	Mean ± SD	Mean ± SD
IL-1B (pg/ml)	1.48 ± 1.21	3.13 ± 6.36	2.33 ± 2.52
IL-4 (pg/ml)	4.97 ± 5.05	2.75 ± 3.58	18.78 ± 26.36
IL-6 (pg/ml)	1237.38 ± 1663.95	2382.74 ± 2957.70	2175.33 ± 2474.16
IL-12 (pg/ml)	1.61 ± 1.65	0.48 ± 0.38	5.92 ± 9.99
IL-17 (pg/ml)	1.41 ± 0.79	4.33 ± 6.04	5.19 ± 4.80
TNFα (pg/ml)	5.99 ± 8.44	16.47 ± 18.59	22.21 ± 22.44
IFNγ (pg/ml)	4.54 ± 4.63	0.59 ± 0.90	8.63 ± 10.72
CRP (mg/dl)	21.91 ± 14.91	22.02 ± 9.60	16.47 ± 11.38
SOFA (score)	10.40 ± 4.30	13.17 ± 1.33	16.70 ± 3.38

Group 1, no catecholamines; Group 2, catecholamines administered, with decreases in their doses noted; and Group 3, catecholamines administered, with stable or increased doses
IL interleukin, *TNFα* tumor necrosis factor alpha, *IFNγ* interferon gamma, *CRP* C-reactive protein, *SOFA* Sequential Organ Failure Assessment

showed an independent contribution of this score to the prediction of change in catecholamines ($r = 0.43$, $R^2 = 0.19$, $p = 0.02$). ROC analysis showed that SOFA score ≤14.0 enabled the prediction of a reduction in catecholamines during HCO-CVVHD with the specificity and sensitivity of 83% (area under curve (AUC) = 0.899, $p < 0.001$) (Fig. 1).

There was a significant correlation between the patient group category, based on catecholamine administration, and the initial level of the cytokines IL-4, IL-12, IL-17, and TNFα and of the SOFA score ($r = 0.42$, $p = 0.008$; $r = 0.33$, $p = 0.043$; $r = 0.43$, $p = 0.006$; $r = 0.39$, $p = 0.016$; $r = 0.68$, $p < 0.001$, respectively). The level of cytokines and SOFA score increased with the consecutive group category (Table 3).

In the multivariate regression analysis, only IL-17, out of the initially correlating cytokines, was independently associated with the catecholamine group category ($R^2 = 0.13$; $p = 0.027$). Moreover, univariate regression analysis showed

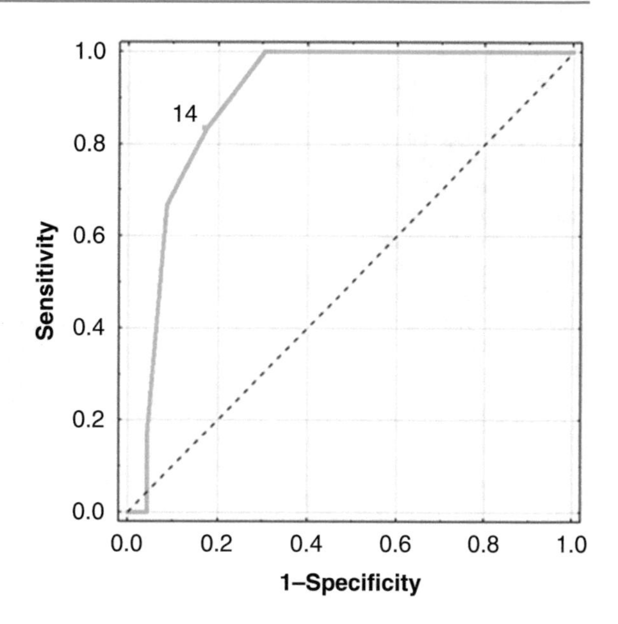

Fig. 1 ROC plot depicting the threshold value of the Sequential Organ Failure Assessment (SOFA) for prediction of a decrease in the use of catecholamines during high cutoff continuous veno-venous hemodialysis (HCO-CVVHD)

a significant correlation between SOFA score and the catecholamine group category change ($R^2 = 0.52$; $p < 0.001$).

The ROC plots showed that the level of IL-17 $\leq$ 1.89 (71.4% sensitivity and 80.0% specificity; AUC = 0.771; $p < 0.001$) and the SOFA scale $\leq$11.0 (100% sensitivity and 70% specificity; AUC = 0.843; $p < 0.001$) enabled the prediction of a necessity of catecholamine treatment during HCO-CVVHD, without a significant difference between either predictive factor ($p = 0.556$) (Figs. 2, 3 and 4). However, the immuno-clinical combination of SOFA $\times$ IL-17 $\leq$ 22.28 pg/mL appeared the strongest predictive factor for treatment with catecholamines during HCO-CVVHD, with 82.1% sensitivity and 90.0% specificity (AUC = 0.843, 95%CI 0.716–0.969; $p < 0.001$) (Fig. 5).

4 Discussion

The major finding of this study was that the initial activation of the immune system, based on IL-17 blood content, was a significant predictor of catecholamine treatment during HCO-CVVHD. The predictive value of the IL-17 marker was on a par with that of the SOFA scale. Moreover, combining these two markers optimized the predictive strength of the anticipation of catecholamine use, with 82% sensitivity and 90% specificity.

Sepsis is the most important cause of acute kidney injury in intensive care units. The injury affects nearly 50% of critically ill septic patients (Kellum and Prowle 2018). Approximately 15–20% of such patients require renal replacement therapy (Peters et al. 2018), with a mortality rate of 45–60% (Harrison et al. 2006; Uchino et al. 2005). During sepsis, severe metabolic disturbances occur at the cellular level. One of the clinical sequalae of sepsis is persistent hypotension requiring administration of vasoconstrictor drugs, such as catecholamines, to stabilize the mean arterial pressure at $\geq$65 mmHg to maintain a proper tissue perfusion. The most frequently recommended and used vasopressor amine is norepinephrine (Rhodes et al. 2017).

A high risk of death in sepsis is associated with the presence of a severe inflammatory reaction, reflected by enhanced serum content of proinflammatory cytokines. Recently, the extracorporeal blood purification from proinflammatory cytokines, using HCO membranes, has gained recognition as a way of suppressing inflammatory reactions in sepsis. The

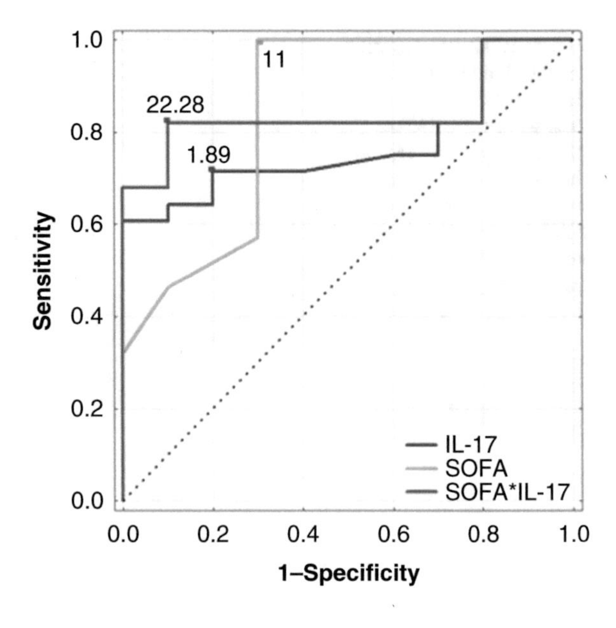

Fig. 2 ROC plots depicting the threshold values of interleukin-17 (IL-17), the Sequential Organ Failure Assessment (SOFA) score, and a combination of the two factors for prediction of decreases in the use of catecholamines during high cutoff continuous veno-venous hemodialysis (HCO-CVVHD)

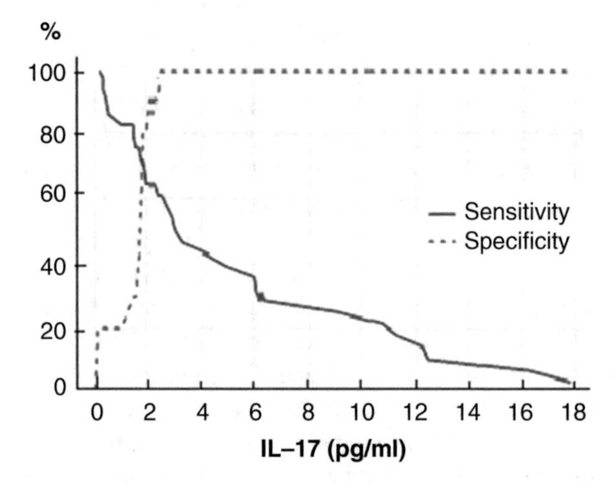

Fig. 3 Sensitivity and specificity of interleukin-17 (IL-17) for prediction of catecholamine treatment during high cutoff continuous veno-venous hemodialysis (HCO-CVVHD)

basis for this type of treatment has been the "maximum concentration" hypothesis. It assumes that avoiding the occurrence of an early peak of circulating proinflammatory molecules may mitigate inflammatory reactions in sepsis. The corollary is the need for arterial blood pressure supportive therapy to maintain organ perfusion (De Vriese et al. 1999).

Studies on the elimination of proinflammatory cytokines are rare and have focused on IL-1B, IL-6, IL-8, IL-18, interleukin-1 receptor antagonist (IL-1RA) and TNFα (Romagnoli et al. 2018; Kade et al. 2016; Mariano et al. 2005; Uchino et al. 2002). Most work has concerned the use of HCO membranes in hemofiltration procedures, in which the convection plays an important role in removing cytokines from the bloodstream. However, a weak point using HCO membranes is a lack of reports on the correlation of cytokine content decrease with clinical patient status (Villa et al. 2014). In the present study, we used hemodialysis treatment, eliminating the risk

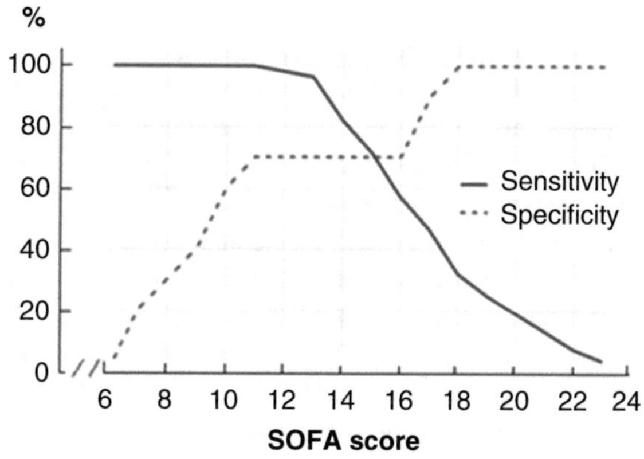

Fig. 4 Sensitivity and specificity of the Sequential Organ Failure Assessment (SOFA) scale for prediction of catecholamine treatment during high cutoff continuous veno-venous hemodialysis (HCO-CVVHD)

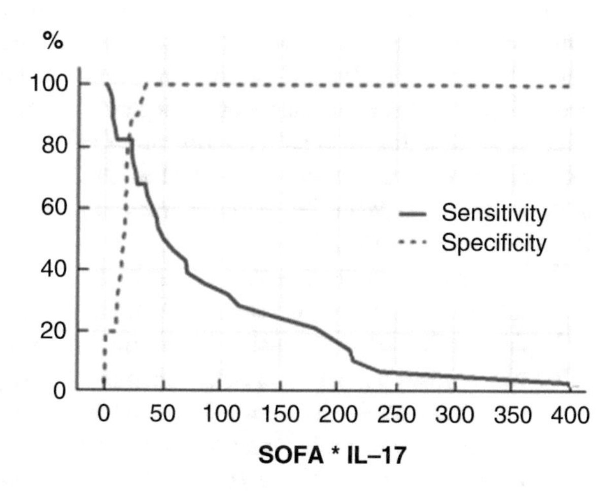

Fig. 5 Sensitivity and specificity of a combination of the Sequential Organ Failure Assessment (SOFA) scale and interleukin-17 (SOFA ∗IL-17) for prediction of catecholamine treatment during high cutoff continuous veno-venous hemodialysis (HCO-CVVHD)

of albumin loss, which has hampered some other studies (Morgera et al. 2004). We also evaluated the clinical condition of patients using the SOFA scale. A low initial level of SOFA score correlated with a lower initial content of IL-4, IL-12, IL-17, and IFNγ. The SOFA scale appeared a good clinical tool to predict who of the patients might require the initiation of catecholamine treatment and in whom catecholamines might possibly be reduced over the course of HCO-CVVHD. In previous clinical reports concerning the use of HCO membranes, the scale of Acute Physiology and Chronic Health Evaluation (APACHE) II and the Scale for the Assessment of Positive Symptoms (SAPS) have been used (Morgera et al. 2004; Morgera et al. 2006). Those scales have also shown a significant improvement in the patients' condition. A reduction in organ systems dysfunction is mainly been associated with improved hemodynamic parameters. That leads to decreased demand for the use of vasopressor agents, which may pointedly be exemplified by significant reductions in the dose of norepinephrine in high-volume hemofiltration therapy using HCO membranes in the studies of Haase et al. (2007) and Morgera et al. (2006).

In the present study, administration of catecholamines was reduced in 6 patients, increased in 5, and remained stable in 17 patients. Norepinephrine was the catecholamine of choice in all but on patient. Further, we showed that lower initial levels of IL-17 and SOFA score were remarkable predictors of circulatory stability, possibly obviating the need for catecholamines in the course of HCO-CVVHD. Lower initial levels of cytokines and SOFA score correlated significantly with reductions in catecholamine doses. In previous reports, hemodynamic changes during hemofiltration using HCO membranes were usually investigated with more invasive methods, such as the measurements of cardiac output, systemic vascular resistance, or pulmonary artery occlusion pressure. Cardiac output has been reported significantly higher and systemic vascular resistance lower during when compared with the respective initial levels, with other features such as requirement for catecholamines or the volume of colloid administered remaining constant throughout hemofiltration. Critically ill patients in septic shock show a growing demand for norepinephrine, which is an indication of clinical deterioration. On the other hand, a continuous downtrend in this demand is a predictor of possible recovery (De Vriese et al. 1999).

IL-17 has emerged as a central player in the mammalian immune system. This cytokine not only plays an important defensive role in many infectious diseases but may also cause inflammation in autoimmune diseases through the induction of IL-6, a major proinflammatory cytokine (Onishi and Gaffen 2010). We have previously shown that excessive activation of the immune system, manifested by an increase in IL-6, has a negative prognostic significance during HCO-CVVHD (Kade et al. 2016). Likewise, CRP, a large pentamer molecule that cannot be removed by a hemofilter, is more useful than procalcitonin in monitoring the inflammation in the course of sepsis in patients undergoing CVVHD (Kade et al. 2018). The serum content of CRP is less influenced by HCO, and thus it could be helpful in monitoring the course of sepsis during extracorporeal blood purification procedures using HCO hemofilters (Caldini et al. 2013). However, in the present work, we failed to substantiate the presence of a significant association of the initial CRP level with the content of the cytokines investigated, the SOFA scale, or catecholamine administration. In this respect, the CRP evaluation appears a rather insensitive method.

This study had some limitations. A group of patients with acute kidney injury was rather small, and the maximum duration of the HCO-CVVHD treatment was just 24 h, which did not enable a longer assessment of catecholamine doses. Nor could HCO-CVVHD treatment be related to the variable content of specific cytokines. Patients received mainly norepinephrine, but there were also patients who required different catecholamines. The number of patients in these subgroups was so small that it prevented from conducting a more reliable intragroup evaluation. The use of HCO-CVVHD remains an experimental method of treatment. The literature reports on the subject matter are still scarce, and most work focuses rather on the use of HCO filters during hemofiltration.

The use of extracorporeal blood purification techniques during sepsis remains a contentious issue, which constrains the efforts to develop guidelines. Hard data remain insufficient to precisely identify septic patients who would benefit most from blood purification procedures. The questions regarding the optimum time to start such treatment and duration and necessary number of treatment repetition remain unanswered as well. Nonetheless, the HCO-CVVHD procedure shows a strong potential to become a valuable form of renal replacement therapy for patients with sepsis complicated by acute kidney injury.

In conclusion, initial activation of the immune system has a significant impact on hemodynamic stability and the need for therapy with catecholamines in patients with septic shock and acute kidney injury during continuous venovenous hemodialysis procedures using high-cutoff hemofilters. The evaluation of both immune status and clinical condition on SOFA scale could be useful in predicting the use of catecholamines and the probability of dosing reduction over the course of treatment.

Acknowledgments Supported by grant no. 179 from the Ministry of Science and Higher Education in Poland.

Conflicts of Interest The authors declare no conflicts of interest in relation to this article.

Ethical Approval All procedures performed in studies involving human participants were in accordance with the ethical standards of the institutional and/or national research committee and with the 1964 Helsinki declaration and its later amendments or comparable ethical standards. The study protocol received approval of the Bioethics Committee of the Military Institute of Medicine in Warsaw, Poland.

Informed Consent Written informed consent was obtained from all individual participants included in the study.

References

Caldini A, Chelazzi C, Terreni A, Biagioli T, Giannoni C, Villa G, Messeri G, De Gaudio AR (2013) Is procalcitonin a reliable marker of sepsis in critically ill septic patients undergoing continuous veno–venous hemodiafiltration with 'high cut–off' membranes (HCO–CVVHDF)? Clin Chem Lab Med 51 (11):261–263

De Vriese AS, Colardyn FA, Philippé JJ, Vanholder RC, De Sutter JH, Lameire NH (1999) Cytokine removal during continuous hemofiltration in septic patients. J Am Soc Nephrol 10(4):846–853

Gierek D, Kuczera M, Dabek J, Piłat D, Kurtok–Nowak A (2011) Results of severe sepsis treatment—two years of experience in a single centre. Anestezjol Intens Ter 43(1):22–28

Haase M, Bellomo R, Baldwin I, Haase–Fielitz A, Fealy N, Davenport P, Morgera S, Goehl H, Storr M, Boyce N, Neumayer HH (2007) Hemodialysis membrane with a high–molecular–weight cutoff and cytokine levels in sepsis complicated by acute renal failure: a phase 1 randomized trial. Am J Kidney Dis 50 (2):296–304

Harrison DA, Welch CA, Eddleston JM (2006) The epidemiology of severe sepsis in England, Wales and Northern Ireland, 1996 to 2004: secondary analysis of a high quality clinical database, the ICNARC case mix programme database. Crit Care 10(2):R42

Kade G, Lubas A, Rzeszotarska A, Korsak J, Niemczyk S (2016) Effectiveness of high cut–off hemofilters in the removal of selected cytokines in patients during septic shock accompanied by acute kidney injury–preliminary study. Med Sci Monit 22:4338–4344

Kade G, Literacki S, Rzeszotarska A, Niemczyk S, Lubas A (2018) Removal of procalcitonin and selected cytokines during continuous veno–venous hemodialysis using high cutoff hemofilters in patients with sepsis and acute kidney injury. Blood Purif 46(2):153–159

Kellum JA, Prowle JR (2018) Paradigms of acute kidney injury in the intensive care setting. Nat Rev Nephrol 14 (4):217–230

Mariano F, Fonsato V, Lanfranco G, Pohlmeier R, Ronco C, Triolo G, Camussi G, Tetta C, Passlick–Deetjen J (2005) Tailoring high–cut–off membranes and feasible application in sepsis–associated acute renal failure: in vitro studies. Nephrol Dial Transplant 20(6):1116–1126

Martin–Loeches I, Guia MC, Vallecoccia MS, Suarez D, Ibarz M, Irazabal M, Ferrer R, Artigas A (2019) Correction to: risk factors for mortality in elderly and very elderly critically ill patients with sepsis: a prospective, observational, multicenter cohort study. Ann Intensive Care 9(1):36

Morgera S, Slowinski T, Melzer C, Sobottke V, Vargas–Hein O, Volk T, Zuckermann–Becker H, Wegner B, Müller JM, Baumann G, Kox WJ, Bellomo R, Neumayer HH (2004) Renal replacement therapy with high–cutoff hemofilters: impact of convection and diffusion on cytokine clearances and protein status. Am J Kidney Dis 43(3):444–453

Morgera S, Haase M, Kuss T, Vargas–Hein O, Zuckermann–Becker H, Melzer C, Krieg H, Wegner B, Bellomo R, Neumayer HH (2006) Pilot study on the effects of high cutoff hemofiltration on the need for norepinephrine in septic patients with acute renal failure. Crit Care Med 34(8):2099–2104

Onishi RM, Gaffen SL (2010) Interleukin–17 and its target genes: mechanisms of interleukin–17 function in disease. Immunology 129(3):311–321

Peters E, Antonelli M, Wittebole X, Nanchal R, François B, Sakr Y, Vincent JL, Pickkers P (2018) A worldwide multicentre evaluation of the influence of deterioration or improvement of acute kidney injury on clinical outcome in critically ill patients with and without sepsis at ICU admission: results from The Intensive Care Over Nations audit. Crit Care 22(1):188

Rhodes A, Evans LE, Alhazzani W et al (2017) Surviving sepsis campaign: international guidelines for management of sepsis and septic shock: 2016. Intensive Care Med 45(3):486–552

Romagnoli S, Ricci Z, Ronco C (2018) CRRT for sepsis–induced acute kidney injury. Curr Opin Crit Care 24 (6):483–492

Uchino S, Bellomo R, Morimatsu H, Goldsmith D, Davenport P, Cole L, Baldwin I, Panagiotopoulos S, Tipping P, Morgera S, Neumayer HH, Goehl H (2002) Cytokine dialysis: an *ex vivo* study. ASAIO J 48 (6):650–653

Uchino S, Kellum JA, Bellomo R, Doig GS, Morimatsu H, Morgera S, Schetz M, Tan I, Bouman C, Macedo E,

Gibney N, Tolwani A, Ronco C, Beginning and Ending Supportive Therapy for the Kidney (BEST Kidney) Investigators (2005) Acute renal failure in critically ill patients: a multinational, multicenter study. JAMA 294 (7):813–818

Villa G, Zaragoza JJ, Sharma A, Neri M, De Gaudio AR, Ronco C (2014) Cytokine removal with high cut–off membrane: review of literature. Blood Purif 38 (3–4):167–173

Xing Z, Gauldie J, Cox G, Baumann H, Jordana M, Lei XF, Achong MK (1998) IL–6 is an antiinflammatory cytokine required for controlling local or systemic acute inflammatory responses. J Clin Invest 101 (2):311–320

Adv Exp Med Biol - Clinical and Experimental Biomedicine (2020) 8: 81–89
https://doi.org/10.1007/5584_2019_444

Published online: 20 November 2019

Body Composition and Biochemical Markers of Nutrition in Non-dialysis-Dependent Chronic Kidney Disease Patients

Aleksandra Rymarz, Maria Zajbt, Anna Jeznach-Steinhagen, Agnieszka Woźniak-Kosek, and Stanisław Niemczyk

Abstract

The aim of this study was to examine the body composition in stages 3b to 5 of chronic kidney disease. There were 149 patients included in the study, with the mean age of 65.5 ± 16.5 years, body mass index (BMI) of 29.4 ± 5.6 kg/m^2, and estimated glomerular filtration rate (eGFR) of 23.2 ± 9.3/min/1.73m^2. They remained with dialysis. Body composition was measured using bioimpedance spectroscopy, and handgrip strength was measured with a hydraulic dynamometer. The main biochemical markers assessed consisted of serum protein, albumin, prealbumin, high-sensitivity C-reactive protein (hsCRP), and interleukin (IL)-6 content. We found that 39% of patients were overweight and 41% were obese. Obesity was more prevalent in stage 3b of chronic kidney disease than in stages 4–5 in women and in patients older than 60 years of age. Thirty-eight percent of the study population were sarcopenic, of whom 20% presented a sarcopenic obesity phenotype. There were significant associations between lean tissue index (LTI) and serum prealbumin content and handgrip strength. Fat tissue index (FTI) was associated and hsCRP, serum protein, body mass index (BMI), waist-hip ratio, and waist-to-height ratio. There were inverse associations between FTI-LTI and LTI-age. We conclude that the prevalence of obesity in non-dialysis-dependent patients with chronic kidney disease is higher than that in the general population. Earlier stages of chronic kidney disease are associated with a higher prevalence of obesity.

Keywords

Biochemical markers · Body composition · Chronic kidney disease · Nutrition · Obesity · Sarcopenia sarcopenic obestity

1 Introduction

Overweight and obesity are a growing problem in the developed countries, with severe health consequences such as cardiovascular disease, diabetes mellitus, hypertension, strokes, and certain types of cancer. According to the WHO registries, the worldwide prevalence of overweight in adults

A. Rymarz (✉), M. Zajbt, and S. Niemczyk
Department of Internal Diseases, Nephrology and Dialysis, Military Institute of Medicine, Warsaw, Poland
e-mail: ola@rymarz.pl; arymarz@wim.mil.pl

A. Jeznach-Steinhagen
Faculty of Health Sciences, Dietetics Division, Warsaw Medical University, Warsaw, Poland

A. Woźniak-Kosek
Department of Laboratory Diagnostics, Military Institute of Medicine, Warsaw, Poland

was 39% in 2016, and it continues to rise. The number of obese individuals doubled between 1980 and 2008 (WHO – Obesity 2019). Body mass index (BMI) $\geq$ 25 kg/m^2 defines the presence of overweight, and $\geq$ 30 kg/m^2 defines obesity. In the general population, obesity is associated with increased mortality (Beleigoli et al. 2012). However, in some groups of patients, this association is the opposite. Populations where the mortality rate decreases as BMI goes up are patients with chronic kidney disease (CKD), obstructive lung disease, chronic heart failure, cancer, and elderly people (Bosello et al. 2016; Pinho et al. 2015). Recent studies have focused on explaining this phenomenon which is referred to as inverse epidemiology or the obesity-survival paradox. One explanation may be that BMI does not differentiate fat mass from lean tissue mass. Therefore, survival analysis based on BMI can be disrupted by a different body composition in individuals with the same BMI. In elderly people and some patients with chronic diseases, muscle mass decreases over time, and a reduction in BMI is not equivalent to a decrease in fat (Cruz-Jentoft et al. 2014).

A reduction in muscle mass and/or muscle strength is called sarcopenia (Cruz-Jentoft et al. 2010). A gradual loss in muscle mass and strength is observed along the aging process (Sakuma and Yamaguchi 2012). Some research distinguishes the term sarcopenia, which refers to the age-related changes, from myopenia, which describes changes induced by pathological processes resulting from diseases. In practice, however, it is hard to distinguish the two conditions, so that the terms are most often used interchangeably (Laviano et al. 2014). The diagnosis of sarcopenia or myopenia is important because a loss of muscle mass and its function is associated with increased mortality in the elderly or in individuals who suffer from specific diseases such as chronic kidney disease (CKD) and certain types of cancer or in patients who underwent cardiac or spinal surgery (Matsuo et al. 2018; Moskven et al. 2018; Okamura et al. 2018). In CKD, a "uremic milieu" enhances a decrease in muscle mass related to age. The mechanisms that underlie sarcopenia in CKD patients are metabolic acidosis, inflammation, hormonal disorders, uremic toxins, oxidative stress, insulin resistance, vitamin D deficiency, and a protein-restricted diet. These mechanisms intensify in advanced stages of CKD. The worst-case scenario for body composition changes is when a decrease in lean tissue mass and its main component, the muscle mass, is accompanied by an increase in fat mass. This kind of phenotype is called sarcopenic obesity. It often accompanies aging, and it is associated with increased mortality in older men (Atkins et al. 2014).

The aim of this study was to examine body composition and biochemical markers of nutrition and inflammation in patients with CKD in stages 3b–5, who were non-dialysis dependent.

2 Methods

The study included 149 patients of the mean age 65.5 $\pm$ 16.5 years with CKD in stages 3b–5 and estimated glomerular filtration rate (eGFR) $<$45 ml/min/1.73m^2. Eighty-two (55%) out of the 149 patients were men. The patients were under a nephrologists' care in the outpatient clinic. The exclusion criteria were clinical signs of fluid overload, infection, chronic heart failure, chronic liver or pulmonary failure, current malignancy, implants containing a metal such as heart pacemakers, stents, or metal stitches, and limb amputations. Bioimpedance spectroscopy was performed using a body composition monitor (Fresenius Medical Care, Shanghai, China). Measurements were taken in the supine position, after a 5-min rest. Electrodes were placed on one hand and one foot in the tetrapolar configuration. Bioimpedance spectroscopy data consisted of the following: lean tissue index defined as a quotient of lean mass over height squared (LTI), fat tissue index defined as a quotient of fat mass over height squared (FTI), and the body cell mass. Handgrip strength was measured using a Saehan hydraulic dynamometer (Masan, South Korea). During the examination, the patient remained seated with the elbow flexed at a 90° angle and the forearm in a neutral position. In each individual, measurements were performed three times at 30 s intervals. The mean of the 3 was taken as a final result.

Blood samples were taken after an overnight fast, and plasma was separated within 30 min. Interleukin 6 (IL-6) was measured using a solid-phase ELISA (Quantikine HS ELISA; R&D Systems, Abingdon, UK). High-sensitivity C-reactive protein (hsCRP) was measured using a nephelometry assay (BNII Siemens Healthcare GmbH, Erlangen, Germany) with a cut-off point of 0.8 mg/dl. Insulin-like growth factor-1 was measured by ELISA (DiAsource ImmunoAssays SA; Louvain-la-Neuve, Belgium). Serum creatinine (SCr) and serum protein (SP), albumin (bromocresol purple), and prealbumin (SPA) were measured using routine clinical methods.

Continuous data were presented as means ± SD and categorical data as percentages. Differences between the groups (men and women and patients above and below 60 years of age) were evaluated using a two-sided independent *t*-test. Correlations between the body compartments and nutritional markers were evaluated with Pearson's correlation coefficient (r). A p-value <0.05 defined statistically significant differences. The analysis was performed using a commercial SPSS v18 package (IBM Corp., Armonk, NY).

Table 1 Clinical characteristic of in non-dialysis-dependent patients with chronic kidney disease (CKD) ($n = 149$)

Parameter	
Age, years	65.5 ± 16.5
Gender, men	82 (55.0%)
SCr, mg/dl	3.1 ± 1.6
eGFR, ml/min/1.73m^2	23.2 ± 9.3
BMI, kg/m^2	29.4 ± 5.6
Waist circumference, cm	101.9 ± 14.2
Hip circumference, cm	107.4 ± 11.4
WHR	0.1 ± 0.1
WHeiR	0.6 ± 0.1
FTI, kg/m^2	16.4 ± 6.1
LTI, kg/m^2	12.5 ± 2.8
BCM, kg	18.6 ± 6.2
Handgrip strength, kg	24.7 ± 14.0
SP, g/dl	6.84 ± 0.76
SA, g/dl	4.12 ± 0.33
SPA, g/dl	32.89 ± 8.34
hsCRP, mg/dl	0.62 ± 0.72
IL-6, pg/ml	7.65 ± 11.22

Data are means ±SD. *SCr* serum creatinine level, *eGFR* estimated glomerular filtration rate, *BMI* body mass index, *WHR* waist-to-hip ratio, *WHeiR* waist-to-height ratio, *FTI* fat tissue index, *LTI* lean tissue index, *BCM* body cell mass, *SP* serum protein level, *SA* serum albumin level, *SPA* serum prealbumin level, *hsCRP* high-sensitivity C-reactive protein, *IL-6*, interleukin-6

3 Results

Clinical characteristics of the study patients are presented in Table 1. The mean eGFR was 23.2 ± 9.3 ml/min/1.73m^2, and the serum creatinine content was 3.1 ± 1.6 mg/dl. The mean BMI was 29.4 ± 5.6 kg/m^2. Out of the 149 patients, 119 (80%) patients had BMI >25 kg/m^2, and 61 (41%) were obese with BMI >30 kg/m^2. The percentage of obese women was higher than that of men (54.4% vs. 36.6%, respectively). The percentage of obese individuals in the subgroups stratified by CKD stages is presented in Table 2. The number of patients with FTI above normal was 52 (34.9%), 2 patients had FTI below the normal range, and the remaining patients had it within the reference range for age and gender. The number of patients with LTI lower than the reference range for age and gender was 57 (38.3%), which indicates a relatively high number of patients presenting a sarcopenic phenotype. Among the patients with LTI lower than the reference range, 31 patients (20.1%) had FTI above normal, which indicates a sarcopenic obesity.

There were significant associations between FTI and BMI ($r = 0.9$, $p = 0.001$), waist-to-hip ratio (WHR) ($r = 0.21$, $p = 0.012$), waist-to-height ratio (WHeiR) ($r = 0.62$, $p < 0.001$), serum protein ($r = 0.21$, $p = 0.013$), and hsCRP ($r = 0.21$, $p = 0.001$). There also were positive correlations between serum prealbumin and LTI ($r = 0.27$, $p = 0.001$) and handgrip strength ($r = 0.42$, $p < 0.001$). Handgrip strength, on the other side, was associated with LTI, body cell mass, waist-to-hip ratio, and serum creatinine content. There were inverse associations between FTI-LTI ($r = -0.36$, $p < 0.001$), FTI-body cell mass ($r = -0.37$, $p < 0.001$), LTI-age ($r = -0.35$, $p < 0.001$), and handgrip strength-age ($r = -0.4$, $p < 0.001$) (Table 3).

The mean BMI was not significantly different between men and woman ($p = 0.562$), nor was it

Table 2 Prevalence of overweight and obesity in non-dialysis-dependent patients with chronic kidney disease (CKD)

Subgroups	BMI > 25 kg/m^2	BMI > 30 kg/m^2
Whole study population	80.0%	40.9%
CKD stage 3b	92.0%	52.0%
CKD stage 4	79.2%	39.8%
CKD stage 5	71.4%	35.7%
Women	77.6%	54.4%
Men	81.7%	36.6%
≥ 60 years old	84.9%	44.3%
< 60 years old	62.8%	32.6%

Table 3 Associations of fat tissue index (FTI), lean tissue index (LTI), and handgrip strength (HGS) with other parameters measured

	FTI		LTI		HGS	
Parameter	*r*	*p*	*r*	p	*r*	*p*
Age, years	0.15	0.068	**−0.350**	**<0.001**	**−0.40**	**<0.001**
BMI, kg/m^2	**0.90**	**0.001**	0.07	0.413	0.05	0.544
WHR	**0.21**	**0.012**	0.17	0.044	**0.24**	**0.006**
WHeiR	**0.62**	**<0.001**	−0.10	0.246	−0.08	0.338
FTI, kg/m^2	–	–	**−0.36**	**<0.001**	−0.09	0.278
LTI, kg/m^2	**−0.36**	**<0.001**	–	–	**0.33**	**<0.001**
BCM, kg	**−0.37**	**<0.001**	**0.95**	**<0.001**	**0.50**	**<0.001**
Handgrip strength, kg	−0.09	0.278	**0.33**	**<0.001**	–	–
SCr, mg/dl	**−0.29**	**<0.001**	**0.25**	**0.002**	**0.39**	**<0.001**
eGFR, ml/min/1.73m^2	**0.20**	**0.013**	−0.15	0.065	−0,08	0.387
SP, g/dl	0.21	0.011	−0.12	0.154	−0.07	0.441
SA, g/dl	−0.15	0.075	0.05	0.576	−0.01	0.882
SPA, g/dl	−0.18	0.027	**0.27**	**0.001**	**0.42**	**<0.001**
hsCRP, mg/dl	**0.21**	**0.001**	0.05	0.560	−0.01	0.896
IL-6, pg/ml	0.15	0.086	−0.08	0.326	−0.08	0.387

BMI body mass index, *WHR* waist to hip ratio, *WHeiR* waist to height ratio, *BCM* body cell mass, *SCr* serum creatinine level, *eGFR* estimated glomerular filtration rate, *SP* serum protein, *SA* serum albumin, *SPA* serum prealbumin level, *hsCRP* high-sensitivity C-reactive protein, *IL-6* interleukin-6. Significant associations are marked in bold

between the patients above and below 60 years of age ($p = 0.886$). However, women had significantly lower LTI, body cell mass, handgrip strength, and the level of prealbumin. The waist circumference and waist-to-hip ratio were significantly larger in men. The patients older than 60 years of age had significantly lower LTI, body cell mass, handgrip strength, and the level of prealbumin. Those below 60 years of age had a significantly greater serum creatinine content and a lower eGFR. Comparative characteristics of the study population by age subgroups are presented in Tables 4 and 5.

4 Discussion

This study revealed a high prevalence of overweight and obesity in non-dialysis-dependent patients in stages 3b–5 of CKD in Poland. The proportion of patients with BMI above 25 kg/m^2 was almost 80%, and those above 30 kg/m^2 were 40.9%. A high prevalence of obesity in CKD patients has also been noticed in other studies. Dierkes et al. (2018) have reported a 53% prevalence of obesity in non-dialysis-dependent patients with CKD stages 3–5 in Norway.

Table 4 Comparative characteristics of non-dialysis-dependent patients with chronic kidney disease (CKD) according to gender

Parameter	Women (n = 67)	Men (n = 82)	p
Age, years	65.2 ± 16.7	65.7 ± 16.3	0.858
SCr, mg/dl	2.6 ± 1.1	3.4 ± 1.9	**0.002**
eGFR, ml/min/1.73m^2	23.0 ± 8.6	23.4 ± 9.9	0.798
BMI, kg/m^2	29.7 ± 6.4	29.2 ± 5.0	0.562
Waist circumference, cm	97.7 ± 15.1	105.3 ± 12.5	**0.001**
Hip circumference, cm	108.5 ± 13.3	106.5 ± 9.5	0.275
WHR	0.9 ± 0.1	1.0 ± 0.1	**<0.001**
WHeiR	0.6 ± 0.1	0.6 ± 0.1	0.371
FTI, kg/m^2	18.0 ± 6.7	15.1 ± 5.3	**0.003**
LTI, kg/m^2	11.4 ± 2.6	13.4 ± 2.6	**<0.001**
BCM, kg	15.0 ± 4.6	21.5 ± 5.8	**<0.001**
Handgrip strength, kg	16.4 ± 7.9	31.6 ± 14.2	**<0.001**
SP, g/dl	6.94 ± 1.00	6.76 ± 0.47	0.161
SA, g/dl	4.14 ± 0.33	4.10 ± 0.33	0.462
SPA, g/dl	31.17 ± 7.00	34.30 ± 9.10	**0.023**
hsCRP, mg/dl	0.62 ± 0.61	0.62 ± 0.81	0.993
IL-6, pg/ml	8.84 ± 14.45	6.70 ± 7.75	0.271

Data are means ±SD. *SCr* serum creatinine level, *eGFR* estimated glomerular filtration rate, *BMI* body mass index, *WHR* waist-to-hip ratio, *WHeiR* waist-to-height ratio, *FTI* fat tissue index, *LTI* lean tissue index, *BCM* body cell mass, *SP* serum protein level, *SA* serum albumin level, *SPA* serum prealbumin level, *hsCRP* high-sensitivity C-reactive protein, *IL-6* interleukin-6. Significant associations are marked in bold

Table 5 Comparative characteristics of non-dialysis-dependent patients with chronic kidney disease (CKD) according to age

Parameter	≥60-year old (n = 106)	<60-year old (n = 43)	p
SCr, mg/dl	2.7 ± 1.1	4.1 ± 2.3	**<0.001**
eGFR, ml/min/1.73m^2	24.6 ± 9.1	19.2 ± 8.7	**0.001**
BMI, kg/m^2	29.4 ± 4.8	29.5 ± 7.5	0.886
Waist circumference, cm	102.3 ± 12.3	100.8 ± 18.8	0.564
Hip circumference, cm	107.28 ± 9.9	107.9 ± 14.9	0.789
WHR	1.0 ± 0.1	0.9 ± 0.1	0.143
WHeiR	0.6 ± 0.1	0.6 ± 0.1	0.021
FTI, kg/m^2	16.9 ± 5.3	15.0 ± 8.0	0.103
LTI, kg/m^2	11.8 ± 2.5	14.4 ± 2.6	**<0.001**
BCM, kg	16.8 ± 5.1	23.7 ± 6.3	**<0.001**
Handgrip strength, kg	22.0 ± 10.8	32.7 ± 18.8	**<0.001**
SP, g/dl	6.89 ± 0.83	6.71 ± 0.51	0.226
SA, g/dl	4.14 ± 0.32	4.07 ± 0.37	0.283
SPA, g/dl	31.17 ± 7.45	37.67 ± 8.91	**<0.001**
hsCRP, mg/dl	0.60 ± 0.60	0.67 ± 1.00	0.565
IL-6, pg/ml	7.98 ± 11.70	6.77 ± 9.89	0.579

Data are means ±SD. *SCr* serum creatinine level, *eGFR* estimated glomerular filtration rate, *BMI* body mass index, *WHR* waist-to-hip ratio, *WHeiR* waist-to-height ratio, *FTI* fat tissue index, *LTI* lean tissue index, *BCM* body cell mass, *SP* serum protein level, *SA* serum albumin level, *SPA* serum prealbumin level, *hsCRP* high-sensitivity C-reactive protein, *IL-6* interleukin-6. Significant associations are marked in bold

Evangelista et al. (2018) have reported about 46% prevalence of obesity in patients in stage 3b CKD and 38% in those in stage 4–5 CKD in South Korea. The prevalence of obesity reported in both studies aforementioned stands higher than that in the general population. In 2016, a WHO registry found the percentage of obese adults in Europe to be between 21 and 23%, compared to 29% in Australia and Canada and 36% in the USA. The relationship between obesity and chronic kidney disease is complicated. Obesity is a risk factor for the development and progression of CKD. Cardiovascular complications associated with obesity influence not only large vessels, such as the aorta or carotid arteries, but also small ones, such as renal vasculature, resulting in a deterioration of kidney function (Kopple and Feroze 2011; Foster et al. 2008). Glomerular hyperfiltration present in obese patients leads to secondary segmental glomerulosclerosis and impaired renal function (Kambham et al. 2001). Diabetes mellitus type 2, which very often accompanies obesity, additionally enhances cardiovascular complications and renal disease progression. CKD causes metabolic disorders and changes the levels of peptides which regulate energy expenditure, such as ghrelin or leptin (Gunta and Mak 2013). Furthermore, fat excess is associated with an overproduction of proinflammatory cytokines, which facilitate inflammation caused by a chronic kidney disease. The results of the present study are in line with the proinflammatory propensity in CKD as FTI was significantly associated with hsCRP. Interestingly, the prevalence of obesity does not manifest as a simple linear correlation with renal function. Evangelista et al. (2018) have noticed the highest prevalence of general and central obesity in stage 2 CKD and a lower one in stages 3, 4, and 5. The results of the present study also support this trend. We found the highest percentage of 52% of obese patients in stage 3b, followed by 39.5% in stage 4, and 35.7% in stage 5 CKD. However, there were no patients in stage 2 or 3a CKD in this study. Therefore, we cannot conclusively report on the obesity prevalence across all severity stages in CKD. Yet, it is a reasonable presumption that a lower percentage of obese patients in advanced CKD stages could result from dietary restrictions or poor appetite.

The fact that there is a higher prevalence of obesity in CKD patients when compared with patients without kidney failure raises questions about the consequences. Interestingly, and in contrast to the general population, the mortality rate decreases as BMI increases in patients with CKD. This survival paradox is explicable by a better nutritional reserve, having a short-term protective effect through minimizing the adverse metabolic effects of CKD. Further, BMI does not exactly distinguish the elements of body composition, whose changes may be masked by the excess of fat tissue. A decrease in lean tissue mass, and its main component the muscle mass, is a relatively frequent unfavorable element in CKD patients. In the present study, we observed an inverse correlation between lean tissue mass and fat mass. A condition when muscle mass decreases while fat mass goes up leads to frailty and has an impact on physical activity, which potentiates lean tissue mass reduction and fat mass augmentation. This vicious cycle can be broken by enhancing physical activity in CKD. Recent studies have shown the efficacy of physical exercise, such as intradialytic resistance training or aerobics for hemodialysis and non-dialysis-dependent patients (Baria et al. 2014; Mihaescu et al. 2013).

The prevalence of sarcopenia in CKD patients is almost as frequent as obesity, ranging from 20% to 44% (Kim et al. 2014; Rosenberger et al. 2014). In the present study, 38% of patients had LTI lower than the reference range for age and gender. LTI was adversely associated with age, being lower in patients above 60 years of age, which is akin to the findings in the general population. Handgrip strength, a surrogate of muscle function, was also inversely associated with age. Some patients with decreased LTI had increased FTI, which is indicative of a sarcopenic obesity phenotype. The prevalence of this phenotype in the studied population was 20.1%. Lim et al. (2018) have noticed a similar percentage of sarcopenic obesity ($\approx$ 23%), contrarily, in patients without kidney disease in the Korean population. Decreased lean tissue mass appears a mortality risk in hemodialysis patients and in non-dialysis-dependent patients in stages 4–5 of CKD (Rymarz et al. 2018; Vega et al. 2017; Isoyama et al. 2014). Taking this unfavorable outcome into account, an early diagnosis of sarcopenia is essential. Therefore, regular monitoring of body composition in patients with CKD is imperative.

Despite a lack of differences between the mean BMI in men and women, the latter manifested a higher amount of fat mass and a lower amount of lean tissue mass in the present study. The percentage of obese individuals, based on the BMI criterion, also was higher in women than that in men (54.4% vs. 36.6%, respectively). In the population of patients without kidney failure, a higher frequency of obesity and excess of fat mass were noticed also in women (Sotillo et al. 2007). The gender-specific differences in the body composition are well known in the general population and are associated with the sex hormone profile. Testosterone, an anabolic factor, induces muscle protein synthesis and muscle formation, which is conducive to higher muscle mass and lower fat tissue mass in both general and CKD populations (Cobo et al. 2017; Yuki et al. 2013). In CKD patients, as the testosterone level decreases, GFR declines leading to a fall in muscle mass. The effects of estrogens on skeletal muscle are less known, especially in CKD patients. However, estrogens may have a protective effect on muscles exerting an anti-inflammatory action (Anderson et al. 2017). Different outcomes are implied due to differences in the body composition between men and women. Recently, Park et al. (2018) have reported a decrease in mortality rate as BMI and serum creatinine content rise. This finding, present but in hemodialysis men and seen also in the present study, could be attributed to a greater muscle mass in male patients, which is reflected in a higher serum creatinine content, a surrogate of muscle mass in patients with very low residual renal function.

Human body is composed of lean tissue mass and fat mass. Somatic protein is retained in the lean tissue mass, whereas the fat mass is a compartment of energy storage. The serum protein, albumin, and prealbumin are enumerated among the biochemical markers that contribute to these compartments. The diagnostic criteria of protein energy wasting (PEW) established by the International Society of Renal Nutrition and Metabolism include a lowering of serum albumin, prealbumin, and cholesterol levels (Fouque et al. 2008). In the present study, there was a significant association between the serum prealbumin level, on the one side, and the lean tissue mass and handgrip strength, on the other side. In other studies, serum prealbumin also was associated with lean tissue mass in CKD patients (Koyun et al. 2018). Interestingly, in the present study, FTI was associated with the protein level. These data suggest that the prealbumin level is a marker that contributes to the lean tissue mass and protein storage in the body, whereas the protein level is a marker of nourishment associated with fat mass.

In synopsis, in this study we found that the prevalence of obesity in non-dialysis-dependent patients with chronic kidney disease was highest in stage 3b and it was progressively lower in stages 4 and 5 of the disease. It also was higher than that present in the general population. Obesity was more frequent in women than men. The prevalence of sarcopenia and sarcopenic obesity was substantial as well, reaching 38% and 20%, respectively. The serum prealbumin level was associated with muscle mass surrogates, the lean tissue index and handgrip strength. We conclude that the investigation of the body composition of patients with chronic kidney disease is essential in the context of longitudinal outcomes.

Conflicts of Interest The authors declare no conflicts of interest in relation to this article.

Ethical Approval All procedures performed in studies involving human participants were in accordance with the ethical standards of the institutional and/or national research committee and with the 1964 Helsinki declaration and its later amendments or comparable ethical standards. The study protocol was accepted by an institutional ethics committee.

Informed Consent Informed consent was obtained from all individual participants included in the study.

References

Anderson LJ, Liu H, Garcia JM (2017) Sex differences in muscle wasting. Adv Exp Med Biol 1043:153–197

Atkins JL, Whincup PH, Morris RW, Lennon LT, Papacosta O, Wannamethee SG (2014) Sarcopenic obesity and risk of cardiovascular disease and mortality: a population-based cohort study of older men. J Am Geriatr Soc 62:253–260

Baria F, Kamimura MA, Aoike DT, Ammirati A, Rocha ML, de Mello MT, Cuppari L (2014) Randomized controlled trial to evaluate the impact of aerobic exercise on visceral fat in overweight chronic kidney disease patients. Nephrol Dial Transplant 29:857–864

Beleigoli AM, Boersma E, Diniz Mde F, Lima-Costa MF, Ribeiro AL (2012) Overweight and class I obesity are associated with lower 10-year risk of mortality in Brazilian older adults: the Bambuí Cohort Study of Ageing. PLoS One 7(12):e52111

Bosello O, Donataccio MP, Cuzzolaro M (2016) Obesity or obesities? Controversies on the association between body mass index and premature mortality. Eat Weight Disord 21:165–174

Cobo G, Gallar P, Di Gioia C, García Lacalle C, Camacho R, Rodriguez I, Ortega O, Mon C, Vigil A, Lindholm B, Carrero JJ (2017) Hypogonadism associated with muscle atrophy, physical inactivity and ESA hyporesponsiveness in men undergoing haemodialysis. Nefrologia 37(1):54–60

Cruz-Jentoft AJ, Baeyens JP, Bauer JM, Boirie Y, Cederholm T, Landi F, Martin FC, Michel JP, Rolland Y, Schneider SM, Topinková E, Vandewoude M, Zamboni M (2010) Sarcopenia: European consensus on definition and diagnosis: Report of the European Working Group on Sarcopenia in Older People. Age Ageing 39:412–423

Cruz-Jentoft AJ, Landi F, Schneider SM, Zúñiga C, Arai H, Boirie Y, Chen LK, Fielding RA, Martin FC, Michel JP, Sieber C, Stout JR, Studenski SA, Vellas B, Woo J, Zamboni M, Cederholm T (2014) Prevalence of and interventions for sarcopenia in ageing adults: a systematic review. Report of the International Sarcopenia Initiative (EWGSOP and IWGS). Age Ageing 43:748–759

Dierkes J, Dahl H, Lervaag Welland N, Sandnes K, Sæle K, Sekse I, Marti HP (2018) High rates of central obesity and sarcopenia in CKD irrespective of renal replacement therapy – an observational cross-sectional study. BMC Nephrol 19:259

Evangelista LS, Cho WK, Kim Y (2018) Obesity and chronic kidney disease: a population-based study among South Koreans. PLoS One 13(2):e0193559

Foster MC, Hwang SJ, Larson MG, Lichtman JH, Parikh NI, Vasan RS, Levy D, Fox CS (2008) Overweight, obesity, and the development of stage 3 CKD: the Framingham Heart Study. Am J Kidney Dis 52:39–48

Fouque D, Kalantar-Zadeh K, Kopple J, Cano N, Chauveau P, Cuppari L, Franch H, Guarnieri G, Ikizler TA, Kaysen G, Lindholm B, Massy Z, Mitch W, Pineda E, Stenvinkel P, Treviño-Becerra A, Wanner C (2008) A proposed nomenclature and diagnostics criteria for protein-energy wasting in acute and chronic kidney disease. Kidney Int 73:391–398

Gunta SS, Mak RH (2013) Ghrelin and leptin pathophysiology in chronic kidney disease. Pediatr Nephrol 28:611–616

Isoyama N, Qureshi AR, Avesani CM, Lindholm B, Bàràny P, Heimbürger O, Cederholm T, Stenvinkel P, Carrero JJ (2014) Comparative associations of muscle mass and muscle strength with mortality in dialysis patients. Clin J Am Soc Nephrol 9:1720–1728

Kambham N, Markowitz GS, Valeri AM, Lin J, D'Agati VD (2001) Obesity-related glomerulopathy: an emerging epidemic. Kidney Int 59:1498–1509

Kim JK, Choi SR, Choi MJ, Kim SG, Lee YK, Noh JW, Kim HJ, Song YR (2014) Prevalence of and factors associated with sarcopenia in elderly patients with end-stage renal disease. Clin Nutr 33:64–68

Kopple JD, Feroze U (2011) The effect of obesity on chronic kidney disease. J Ren Nutr 21:66–71

Koyun D, Nergizoglu G, Kir KM (2018) Evaluation of the relationship between muscle mass and serum myostatin levels in chronic hemodialysis patients. Saudi J Kidney Dis Transpl 29:809–815

Laviano A, Gori C, Rianda S (2014) Sarcopenia and nutrition. Adv Food Nutr Res 71:101–136

Lim HS, Park YH, Suh K, Yoo MH, Park HK, Kim HJ, Lee JH, Byun DW (2018) Association between sarcopenia, sarcopenic obesity, and chronic disease in Korean elderly. J Bone Metab 25:187–193

Matsuo Y, Mitsuyoshi T, Shintani T, Iizuka Y, Mizowaki T (2018) Impact of low skeletal muscle mass on non-lung cancer mortality after stereotactic body radiotherapy for patients with stage I non-small cell lung cancer. J Geriatr Oncol 9:589–593

Mihaescu A, Avram C, Bob F, Gaita D, Schiller O, Schiller A (2013) Benefits of exercise training during hemodialysis sessions: a prospective cohort study. Nephron Clin Pract 124:72–78

Moskven E, Bourassa-Moreau É, Charest-Morin R, Flexman A, Street J (2018) The impact of frailty and sarcopenia on postoperative outcomes in adult spine surgery. A systematic review of the literature. Spine J 18(12):2354–2369

Okamura H, Kimura N, Tanno K, Mieno M, Matsumoto H, Yamaguchi A, Adachi H (2018) The impact of preoperative sarcopenia, defined based on psoas muscle area, on long-term outcomes of heart valve surgery. J Thorac Cardiovasc Surg. https://doi.org/10.1016/j.jtcvs.2018.06.098

Park JM, Lee JH, Jang HM, Park Y, Kim YS, Kang SW, Yang CW, Kim NH, Kwon E, Kim HJ, Lee JE, Jung HY, Choi JY, Park SH, Kim CD, Cho JH, Kim YL, Clinical Research Center for End Stage Renal Disease (CRC for ESRD) Investigators (2018) Survival in patients on hemodialysis: effect of gender according to body mass index and creatinine. PLoS One 13:e0196550

Pinho EM, Lourenço P, Silva S, Laszczyńska O, Leite AB, Gomes F, Pimenta J, Azevedo A, Bettencourt P (2015) Higher BMI in heart failure patients is associated with longer survival only in the absence of diabetes. J Cardiovasc Med (Hagerstown) 16:576–582

Rosenberger J, Kissova V, Majernikova M, Straussova Z, Boldizsar J (2014) Body composition monitor assessing malnutrition in the hemodialysis population independently predicts mortality. J Ren Nutr 24:172–176

Rymarz A, Gibińska J, Zajbt M, Piechota W, Niemczyk S (2018) Low lean tissue mass can be a predictor of one-year survival in hemodialysis patients. Ren Fail 40:231–237

Sakuma K, Yamaguchi A (2012) Novel intriguing strategies attenuating to sarcopenia. J Aging Res 2012:251217

Sotillo C, López-Jurado M, Aranda P, López-Frías M, Sánchez C, Llopis J (2007) Body composition in an adult population in southern Spain: influence of lifestyle factors. Int J Vitam Nutr Res 77:406–414

Vega A, Abad S, Macías N, Aragoncillo I, Santos A, Galán I, Cedeño S, Manuel López-Gómez J (2017) Low lean tissue mass is an independent risk factor for mortality in patients with stages 4 and 5 non-dialysis chronic kidney disease. Clin Kidney J 10:170–175

WHO – Obesity (2019) http://www.who.int/topics/obesity/en/. Accessed on 31 Aug 2019

Yuki A, Otsuka R, Kozakai R, Kitamura I, Okura T, Ando F, Shimokata H (2013) Relationship between low free testosterone levels and loss of muscle mass. Sci Rep 3:1818

Adv Exp Med Biol - Clinical and Experimental Biomedicine (2020) 8: 91–97
https://doi.org/10.1007/5584_2019_461

Published online: 13 December 2019

Biocompatibility of Hemodialysis

Małgorzata Gomółka, Longin Niemczyk, Katarzyna Szamotulska, Magdalena Mossakowska, Jerzy Smoszna, Aleksandra Rymarz, Leszek Pączek, and Stanisław Niemczyk

Abstract

This study was designed to investigate the biocompatibility of hemodialysis procedures, largely depending on the contact of patient's blood with the dialysis membranes. We addressed the issue by comparing the content of the proteolytic enzymes collagenase and cathepsin B and that of neutrophil myeloperoxidase (MPO) and C-reactive protein (CRP) in the blood before and after a single session treatment and a full course of successive 8-week-long therapies with three types of hemodialysis: low-flux (lfHD), high-flux (hfHD), and post-dilution hemodiafiltration (HDF). The study included 19 patients with chronic nephropathy. We found that collagenase significantly increased after a single session of each type of hemodialysis. Cathepsin B tended to decrease after single sessions; the decrease reached significance only after hfHD. CRP increased significantly after single hfHD and HDF treatments. These changes were meager, with no differences depending on the dialysis type, and their significance was lost after 8-week-long therapy, except the persisting increase in CRP after HDF. Neutrophil MPO apparently was not activated during any type of dialysis, as its content was below the detection threshold. We conclude that all three types of hemodialysis are compatible with the biological system, so that they would rather unlikely lead to clinically harmful effects in chronically hemodialyzed patients. Nonetheless, proteolytic enzymes and myeloperoxidase seem hardly appropriable estimators of hemodialysis biocompatibility due to meager and variable changes. Upregulation of C-reactive protein, on the other hand, expresses a general pro-inflammatory propensity of hemodialysis and is not a suitable estimator of biocompatibility either.

Keywords

Biocompatibility · Hemodialysis · Inflammation · Kidney insufficiency · Kidney replacement therapy · Myeloperoxidase · Nephropathy · Proteolytic enzymes

M. Gomółka, M. Mossakowska, J. Smoszna, A. Rymarz, and S. Niemczyk
Department of Internal Medicine, Nephrology and Dialysis, Military Institute of Medicine, Warsaw, Poland

L. Niemczyk (✉)
Department of Nephrology, Dialysis and Internal Medicine, Medical University of Warsaw, Warsaw, Poland
e-mail: lniemczyk@wum.edu.pl

K. Szamotulska
Department of Epidemiology and Biostatistics, Institute of Mother and Child in Warsaw, Warsaw, Poland

L. Pączek
Department of Immunology, Transplantology and Internal Medicine, Medical University of Warsaw, Warsaw, Poland

1 Introduction

Hemodialysis is a procedure performed in the extracorporeal circulation and may not be biocompatible, which means that it can evoke clinically significant responses, mainly activation of inflammatory and immune pathways (Hakim 2000). Biocompatibility problems during hemodialysis stem from the contact of patient's blood with the dialysis membranes, dialyzate and intravenous fluids administered, as well as drains used in the procedure. The quality of a dialysis membrane is particularly important due to its surface and time of contact with blood (Kalantar-Zadeh et al. 2004). Another essential attribute of the dialysis membrane is roughness (porosity) of its surface as it affects the adhesion of platelets to it. The membrane roughness and its structural and chemical texture, along with functional parameters of a dialyzer, molecular mass transfer, and ultrafiltration coefficients, may all adversely influence dialysis biocompatibility by activation of complement and coagulation systems, by pro-inflammatory cytokines, and by stimulation of proteolytic enzymes by granulocytes (Cohen-Mazor et al. 2014; Gritters et al. 2006; Lin et al. 1996; Tsunoda et al. 1999). Studies on the influence of various types of hemodialysis per se on biocompatibility are scarce, and the information is contentious or unsettled despite technological advances of late (Grano et al. 2003; Otsubo et al. 2000). Therefore, the aim of this study was to define the activity of proteolytic enzymes and myeloperoxidase (MPO) and the content of high-specificity C-reactive protein (hsCRP) during the following dialysis conditions:

1. Single session treatment with low-flux hemodialysis (lfHD), high-flux hemodialysis (hfHD), and post-dilution hemodiafiltration (HDF).
2. Eight-week-long treatment with lfHD, hfHD, and HDF.

2 Methods

This article is an extension of the scope of our previously published investigation on the effectiveness of removal of indoxyl sulfate and p-cresol sulfate, the protein-bound uremic toxins that underlie chronic kidney insufficiency, depending on the type of hemodialysis. The present investigation was performed in the same group of patients, using the same protocol of three different types of hemodialysis employed in succession (Gomółka et al. 2019). We herein focused on the influence of hemodialysis on the activities of collagenase, cathepsin B, and neutrophil MPO, as well as on the content of hsCRP. We considered a separate ramification of our research on kidney replacement therapy, covering new ground of biocompatibility of hemodialysis. The study was completed by 19 patients (7 F, 12 M), including 8 patients (3 F, 5 M) with diabetic nephropathy. The mean age of the 19 patients was 54 ± 15 (range 30–72) years, with body mass index (BMI) 28.6 (19.3–43.0 kg/m^2). They were treated with renal replacement therapy by a mean of 20.0 ± 14.4 (range 3.7–61.2) months. The patients were in a stable condition and were treated three times a week, with Kt/V > 1.2 (K, urea clearance; t, dialysis time; V, urea distribution volume), a dimensionless parameter derived from pretreatment and posttreatment blood urea nitrogen. Each treatment took 237 ± 24 (range 180–270) min. The substitute volume during HDF was 14.1 ± 1.5 (range 10.8–16.2) L per treatment. The average 24-h diuresis amounted to 737 mL, with only two patients having anuria.

Patients with clinically evident inflammation, autoimmune diseases, cancer, malnutrition (BMI <19 kg/m^2), thyroid disorders, severe liver damage, and blood coagulation disorders and on hormonal or oral anticoagulant therapy were excluded. The study protocol lasted for 24 weeks and consisted of three consecutive hemodialysis types, each lasting for 8 weeks: lfHD (low-flux Braun LOPS dialyzers), hfHD (high-flux Braun HIPS dialyzers), and HDF (post-dilution hemodiafiltration; three times a week, with Braun HIPS dialyzers). The effect on the blood indices below outlined of a single session for lfHD, hfHD, and HDF was assessed by collecting blood before and after the first treatment session of each type. The effect of the entire 8-week-long therapy with each type of hemodialysis was taken as a difference in the activity level between onset and end of treatment, the latter marking the beginning level of the next treatment type.

The activities of collagenase and cathepsin B and that of neutrophil MPO were assessed by fluorometry (Cayman Chemicals; Ann Arbor, MI). hsCRP was assessed by nephelometry (BN II analyzer; Siemens, Munich, Germany). Low-molecular heparin, *Dalteparinum natricum*, was administered for blood anticoagulation.

Nonparametric Wilcoxon rank-sum test or the Friedman test was used to compare two or three matched groups over time, respectively. A p-value <0.05 defined statistically significant changes. A commercial IBM SPSS v.24 package (IBM Corp, Armonk, NY) and R v.3.4 software were used.

3 Results

The activity of collagenase significantly rose after a single session of each type of hemodialysis tested; the increases did not differ from one another. Cathepsin B tended to decrease after the use of each type of dialysis; the decrease reached statistical significance and significantly differed from the other two types of hemodialysis only after a single hfHD treatment. The content of hsCRP increased significantly after single hfHD and HDF treatments, with no differences depending on the dialysis type (Table 1). In contrast to single dialysis sessions, there were no significant differences among changes in collagenase, cathepsin B, and hsCRP after the full course of 8-week-long therapy with the three dialysis types, except the persisting increase in hsCRP after HDF (Table 2).

Table 1 Changes in the content of proteolytic enzymes and C-reactive protein after a single session of lfHD, hfHD, and HDF dialysis procedures

Indices	Dialysis type	Before dialysis	After dialysis session	$p_{Wilcoxon}$	Δ	$p_{Δ}$	Δ%	$p_{Δ\%}$
Collagenase median (range)	lfHD	8.5 (5.1–32.4)	9.1 (3.0–38.0)	**0.041**	0.8 (−13.8; +5.6)	0.688	10.3 (−60.6; +41.4)	0.763
	hfHD	7.7 (4.8–30.4)	8.8 (3.8–35.2)	**0.050**	0.6 (−15.1; +4.8)		8.9 (−74.0; +56.9)	
	HDF	7.7 (4.7–38.3)	9.3 (5.1–41.8)	**0.013**	1.2 (−2.9; +3.5)		9.1 (−23.8; +48.5)	
Cathepsin B mean ± SD	lfHD	14.8 ± 3.7	14.7 ± 3.6	0.496	−0.1 ± 1.6	**0.023**	0.1 ± 11.4	**0.023**
	hfHD	15.1 ± 3.6	13.1 ± 3.5	**0.006**	−2.1 ± 3.1		−11.7 ± 21.6	
	HDF	14.7 ± 3.6	13.8 ± 4.8	0.307	−0.9 ± 3.2		−6.9 ± 26.1	
hsCRP median (range)	lfHD	0.59 (0.02–3.52)	0.59 (0.02–3.65)	0.135	0.01 (−0.16; +0.60)	0.126	2.8 (−12.5; +71.4)	0.889
	hfHD	0.72 (0.02–4.37)	0.78 (0.02–4.35)	**0.035**	0.02 (−0.06; +0.29)		2.9 (−7.0; +21.1)	
	HDF	0.43 (0.02–8.37)	0.53 (0.02–7.28)	**0.020**	0.05 (−1.09; +1.06)		10.1 (−14.3; +64.3)	

lfHD low-flux hemodialysis, *hfHD* high-flux hemodialysis, *HDF* post-dilution hemodiafiltration, *hsCRP* high-specificity C-reactive protein, statistically significant changes in bold

Table 2 Changes in the content of proteolytic enzymes and C-reactive protein during 8-week-long cycles of different types of dialysis: lfHD, hfHD, and HDF

Indices	Dialysis type	Onset of observation	End of observation	p	Δ	$p_{Δ}$	Δ%	$p_{Δ\%}$
Collagenase median (range)	lfHD	8.5 (5.1–32.4)	7.7 (4.8–30.4)	0.245	−0.50 (−14.2; +14.2)	0.148	−6.1 (−62.3; +229.0)	0.134
	hfHD	7.7 (4.8–30.4)	7.7 (4.7–38.3)	0.509	0.30 (−15.1; +7.9)		6.3 (−74.0; +55.8)	
	HDF	7.7 (4.7–38.3)	7.4 (3.5–35.6)	0.368	−0.20 (−4.7; +3.4)		−3.2 (−50.0; +51.5)	
Cathepsin B mean ± SD	lfHD	14.76 ± 3.65	15.14 ± 3.65	0.251	0.38 ± 2.05	0.944	3.2 ± 13.3	0.905
	hfHD	15.14 ± 3.65	14.71 ± 3.59	0.390	−0.44 ± 2.90		−0.7 ± 23.1	
	HDF	14.71 ± 3.59	14.27 ± 3.61	0.687	−0.44 ± 3.76		0.4 ± 26.5	
hsCRP median (range)	lfHD	0.59 (0.02–3.52)	0.72 (0.02–4.37)	0.873	−0.04 (−2.80; +4.08)	0.144	−4.9 (−90.0; +1407.0)	0.144
	hfHD	0.72 (0.02–4.37)	0.43 (0.02–8.37)	0.758	0.00 (−3.85; +6.03)		0.0 (−88.0; +323.0)	
	HDF	0.43 (0.02–8.37)	0.65 (0.02–23.60)	**0.028**	0.31 (−7.76; +20.30)		57.5 (−93.0; +650.0)	

lfHD low-flux hemodialysis, *hfHD* high-flux hemodialysis, *HDF* post-dilution hemodiafiltration, *hsCRP* high-specificity C-reactive protein, statistically significant changes in bold

Neutrophil MPO activity was below the threshold of detection by fluorometry in 101 (70%) out of the 141 measurements, which suggests the enzyme was not activated during the hemodialysis procedures employed. Thus, MPO data were excluded from further elaboration.

4 Discussion

This study investigated the biocompatibility of hemodialysis, which mostly depends on the contact of blood flowing through membranes in dialysis devices. The goal of performing biocompatibility testing was to define the fitness of basic types of hemodialysis employed in patients on chronic kidney replacement therapy, concerning the appearance of potentially harmful pathophysiological effects. As a surrogate of biocompatibility, we took changes in the blood content of proteolytic enzymes, myeloperoxidase, and C-reactive protein using low-flux, high-flux hemodialysis, and post-dilution hemodiafiltration, during both the short-term single session treatment and the long-term 8-week-long therapy. The rationale for investigating these enzymes was that they are key players in protein metabolism and oxidative and inflammatory propensity, the pathways most vulnerable to dysregulation during hemodialysis. In general, we found meager changes in the enzyme content, and if there were any they hardly differed and thus were rather independent on the type of hemodialysis. The changes were mostly limited to the single dialysis treatment, when collagenase activity appreciably decreased after all three types of dialysis. Cathepsin B slightly decreased after hfHD and HDE. However, the decreases did not exceed about 10%. Interestingly, these changes flattened out after the full 8-week-long courses of all three types of dialysis, as if there had been some effective counter bio-measures raised to reverse the potentially positive changes concerning biocompatibility over the long-run therapy. The exception was a rise in hsCRP, particularly apparent after HDF, which persisted, albeit at a modest level until therapy completion.

The present findings failed to support the usefulness of myeloperoxidase as a marker of pro-oxidative state. Myeloperoxidase is an enzyme abundantly present in neutrophils and is also prominently linked to inflammation and cardiovascular disorders, frequent adverse effects of dialysis (Aratani 2018; Schindhelm et al. 2009). A great majority of myeloperoxidase measurements in this study rendered null values, below the detection limit. That suggests unsuitability of this enzyme as a marker of biocompatibility of dialysis in terms of pro-inflammatory propensity, which actually confirms the findings by other authors (Grano et al. 2003; Malle et al. 2003), despite the enzyme's documented modulatory role in other inflammatory states, including sepsis or obesity (Savonius et al. 2019; Lauhio et al. 2016; Hou et al. 2013). Wu et al. (2005) have reported a prompt increase in plasma MPO level during the first 15 min of single session dialysis with two different polysulfone and cellulose dialysis membranes. Likewise, Krieter et al. (2006) have found a robust sixfold increase in MPO within the first 5 min of dialysis. In both reports, however, there is no mention on the anticoagulation strategy during hemodialysis. Meanwhile, Borawski (2006) has pointed out to the potential role of heparin, which is commonly used during hemodialysis, in upregulating MPO activity. Heparin is conducive to MPO release from vascular walls; thereby it increases MPO level. In the present study, we used a low-molecular heparin derivative for anticoagulation, which as opposed to high-molecular heparin mitigates the upregulation of MPO (Ono et al. 2000).

Peptidases referred as proteases or proteinases are enzymes which hydrolyze peptide bonds. Depending on the main functional group and reaction mechanism, the enzymes are stratified into serine, cysteine, aspartic, and threonine proteases and metalloproteinases. Cathepsin B, investigated in this study, belongs to cysteine proteases and cathepsin L, and collagenase is part of the family of metalloproteinases (Davies et al. 1992). Changes in cathepsins and collagenases have been differentially reported in various studies on hemodialyzed patients as unchanged when compared with a control group of subjects or increased (Cohen-Mazor et al. 2014; Chou et al. 2002). In the present study, we noticed a general tendency for a decrease in cathepsin B, which assumed significance only concerning hfHD and only during a single procedure. Collagenase, on the other hand, significantly increased during all types of hemodialysis employed, also only during single procedures, and with no appreciable differences among the three procedures. Chou et al. (2002) have reported an increase in tissue inhibitor of metalloproteinases-2, which may explain an increase in some enzymes, and a significant decrease in MMP-9 and MMP-2 activity after hemodialysis. Inose et al. (2000) have reported that the activities of cathepsin B and cathepsin L do not increase, which may also be due to removal of the enzymes, anyhow suggesting their rather low usefulness in the assessment of biocompatibility during hemodialysis. In line with the aforementioned, Otsubo et al. (2000) have reported that the activity of proteolytic enzymes such as elastase, cathepsins, and collagenase in hemodialyzed patients is similar to that in the control non-hemodialyzed. Further, the patients' activity decreases after the in vitro addition of tumor necrosis factor alpha into the assay. A spate of intertwined and interactive biological and technical dependencies and factors during hemodialysis makes it hard and uncertain to sort out changes in the activity of specific proteolytic enzymes or oxidative stress markers during the procedure. That seems essential to an array of discrepant results obtained in various studies on the matter discussed above and also is in line with the present findings putting into doubt the investigation of proteolytic enzyme activity as a measure of hemodialysis biocompability.

The only blood index that more clearly changed in the present study was hsCRP which was significantly upregulated after a single session of hemodialysis, with the strongest increase after HDF. The upregulation abated after full 8-week-long courses of hemodialysis, retaining significance only in case of HDF. A wide spread of hsCRP data made the interpretation of any possible differences stemming from the type of hemodialysis, thereby also in the assessment of biocompatibility of hemodialysis per se, inconclusive. Hemodialysis, generally, is known to be a pro-inflammatory condition. Borazan et al.

(2004) have investigated CRP and pro-inflammatory cytokines in patients with end-stage renal disease on either hemodialysis or peritoneal dialysis and found them increased regardless of dialysis type. Likewise, other previous studies confirm the upregulation of CRP in hemodialyzed patients, which is accompanied by increased risk of cardiovascular and inflammatory disorders and mortality (Pawlak et al. 2005; Zimmermann et al. 1999; Owen and Lowrie 1998; Docci et al. 1990).

In conclusion, the findings of this study demonstrate that all three types of hemodialysis devices in common use are sufficiently compatible with the biological system in that they would, in and of itself, be a rather unlikely cause of any harmful effects or a sustained clinical worsening in chronically hemodialyzed patients, at least within the bio-function of the enzyme tested. Nonetheless, proteolytic enzymes and myeloperoxidase seem hardly appropriable estimators of hemodialysis biocompatibility due to meager changes and variability depending on a multitude of poorly controllable intertwined biofunctional connections. C-reactive protein, on the other hand, is ubiquitously upregulated and expresses a general pro-inflammatory propensity of hemodialysis, regardless of the dialysis type, and as such is not a suitable estimator of biocompatibility either. Further exploration of indices that could offer the optimum assessment of biocompatibility of hemodialysis devices is required using alternative study designs.

Acknowledgments Funded by a grant no 258 from the Military Institute of Medicine.

Conflicts of Interest The authors declare no conflict of interest in relations to this article.

Ethical Approval All procedures performed in studies involving human participants were in accordance with the ethical standards of the institutional and/or national research committee and with the 1964 Helsinki Declaration and its later amendments or comparable ethical standards. This study was approved by the Bioethics Committee of the Military Institute of Medicine in Warsaw, Poland – 05-2013.

Informed Consent Informed consent was obtained from all individual participants included in the study.

References

Aratani Y (2018) Myeloperoxidase: its role for host defense, inflammation, and neutrophil function. Arch Biochem Biophys 640:47–52

Borawski J (2006) Myeloperoxidase as a marker of hemodialysis biocompatibility and oxidative stress: the underestimated modifying effects of heparin. Am J Kidney Dis 47:37–41

Borazan A, Ustun H, Ustundag Y, Aydemir S, Bayraktaroglu T, Sert M, Yilmaz A (2004) The effects of peritoneal dialysis and hemodialysis on serum tumor necrosis factor–alpha, interleukin–6, interleukin–10 and C–reactive–protein levels. Mediat Inflamm 13:201–204

Chou FP, Chu FC, Cheng MC, Yang SF, Cheung WN, Chiou HL, Hsieh YS (2002) Effect of hemodialysis on the plasma level of type IV collagenases and their inhibitors. Clin Biochem 35:383–388

Cohen-Mazor M, Mazor R, Kristral B, Sela S (2014) Elastase and cathepsin G from primed leukocytes cleave vascular endothelial cadherin in hemodialysis patients. Biomed Res Int 2014:459640

Davies M, Martin J, Thomas GJ, Lovett DH (1992) Proteinases and glomerular matrix turnover. Kidney Int 41:671–678

Docci D, Bilancioni R, Buscaroli A, Baldrati L, Capponcini C, Mengozzi S, Turci F, Feletti C (1990) Elevated serum levels of C–reactive protein in hemodialysis patients. Nephron 56:364–367

Gomółka M, Niemczyk L, Szamotulska K, Wyczałkowska-Tomasik A, Rymarz A, Smoszna J, Jasik M, Pączek L, Niemczyk S (2019) Protein-bound solute clearance during hemodialysis. Adv Exp Med Biol 1153:69–77

Grano V, Diano N, Portaccio M, De Santo N, Di Martino S, Rossi S, De Santo LS, Salamino F, Mattei A, Mita DG (2003) Protease removal by means of antiproteases immobilized on supports as a potential tool for hemodialysis or extracorporeal blood circulation. Int J Artif Organs 26:39–45

Gritters M, Grooteman MC, Schoorl M, Schoorl M, Bartels PC, Scheffer PG, Teerlink T, Schalkwijk CG, Spreeuwenberg M, Nube MJ (2006) Citrate anticoagulation abolishes degranulation of polymorphonuclear cells and platelets and reduces oxidative stress during haemodialysis. Nephrol Dial Transplant 21:153–159

Hakim RM (2000) Clinical implications of biocompatibility in blood purification membranes. Nephrol Dial Transplant 15(Suppl 2):16–20

Hou ZH, Lu B, Gao Y, Cao HL, Yu FF, Jing N, Chen X, Cong XF, Roy SK, Budoff MJ (2013) Matrix metalloproteinase-9 (MMP-9) and myeloperoxidase (MPO) levels in patients with nonobstructive coronary artery disease detected by coronary computed tomographic angiography. Acad Radiol 20(1):25–31

Inose K, Ono K, Okubo Y, Yoshida R, Naruse T (2000) The elevation of plasma levels of myeloperoxidase and polymorphonuclear leukocyte elastase as an index of bioincompatibility of the column during hemodialysis using with a β_2–microglobulin–selective adsorbent column. Clin Exp Nephrol 4:52–57

Kalantar-Zadeh K, Block G, CJ MA, Humphreys MH, Kopple JD (2004) Appetite and inflammation, nutrition, anemia, and clinical outcome in hemodialysis patients. Am J Clin Nutr 80:299–307

Krieter DH, Lemke HD, Wanner C (2006) Myeloperoxidase serves as a marker of oxidative stress during single haemodialysis session using two different biocompatible membranes. Nephrol Dial Transplant 21:546

Lauhio A, Farkkila E, Pietilainen KH, Astrom P, Winkelmann A, Tervahartiala T, Pirila E, Rissanen A, Kaprio J, Sorsa TA, Salo T (2016) Association of MMP–8 with obesity, smoking and insulin resistance. Eur J Clin Investig 46:757–765

Lin YF, Chang DM, Shaio MF, Lu KC, Chyr SH, Li BL, Sheih SD (1996) Cytokine production during hemodialysis: effect of dialytic membrane and complement activation. Am J Nephrol 16:293–299

Malle E, Buch T, Grone HJ (2003) Myeloperoxidase in kidney disease. Kidney Int 64:1956–1967

Ono K, Ueki K, Inose K, Tsuchida A, Yano S, Nojima Y (2000) Plasma levels of myeloperoxidase and elastase are differentially regulated by hemodialysis membranes and anticoagulants. Res Commun Mol Pathol Pharmacol 108:341–349

Otsubo H, Kaito K, Shiba K (2000) Flowcytometric analysis on neutrophil intracellular enzyme activity in patients on hemodialysis and continuous ambulatory peritoneal dialysis. Kansenshogaku Zassi 74:73–81

Owen WF, Lowrie EG (1998) C–reactive protein as an outcome predictor for maintenance hemodialysis patients. Kidney Int 54:627–636

Pawlak K, Pawlak D, Myśliwiec M (2005) Circulating beta–chemokines and matrix metalloproteinase–9/tissue inhibitor of metalloproteinase–1 system in hemodialyzed patients – role of oxidative stress. Cytokine 31:18–24

Savonius O, Roine I, Alassiri S, Tervahartiala T, Helve O, Fernández J, Peltola H, Sorsa T, Pelkonen T (2019) The potential role of matrix metalloproteinases 8 and 9 and myeloperoxidase in predicting outcomes of bacterial meningitis of childhood. Mediators Inflamm, Article ID 7436932. https://doi.org/10.1155/2019/7436932

Schindhelm RK, van der Zwan LP, Teerlink T, Scheffer PG (2009) Myeloperoxidase: a useful biomarker for cardiovascular disease risk stratification? Clin Chem 55(8):1462–1470

Tsunoda N, Kokubo K, Sakai K, Fukuda M, Miyazaki M, Hiyoshi T (1999) Surface roughness of cellulose hollow fiber dialysis membranes and platelet adhesion. ASAIO J 45:418–423

Wu CC, Chen JS, Wu WM, Liao TN, Chu P, Lin SH, Chuang CH, Lin YF (2005) Myeloperoxidase serves as a marker of oxidative stress during single haemodialysis session using two different biocompatible dialysis membranes. Nephrol Dial Transplant 20:1134–1139

Zimmermann J, Herrlinger S, Pruy A, Metzger T, Wanner C (1999) Inflammation enhances cardiovascular risk and mortality in hemodialysis patients. Kidney Int 55:648–658

Adv Exp Med Biol - Clinical and Experimental Biomedicine (2020) 8: 99–105
https://doi.org/10.1007/5584_2019_446

Published online: 10 November 2019

Decreasing Vaccination Coverage Against Hepatitis B and Tuberculosis in Newborns

Aneta Nitsch-Osuch, Beata Pawlus, Maria Pawlak, and Ernest Kuchar

Abstract

The number of parents who refuse to vaccinate their children or present the so-called hesitant behavior, i.e., delay the moment of vaccination beyond the mandatory time, has increased in many developed countries. The purpose of this retrospective study was to evaluate the completeness and timeliness of vaccinations against hepatitis B (HBV) and tuberculosis (TB) in neonates in a single maternity hospital in Warsaw, Poland. We reviewed medical files of 14,785 children born in the hospital in 2015–2017 and calculated the proportion of newborns not vaccinated on time according to the Polish Immunization schedule that includes vaccination against HBV and TB in the first day of life. Newborns remained unvaccinated because of parental refusal (refusers) or decision for a delay (hesitants), or medical contraindications. The percentage of unvaccinated newborns in the 3 years was as follows: 7.3% in 2015, 6.7% in 2016, and 10.1% in 2017. Parental decisions rather than medical contraindications caused nonvaccination (4.4% vs. 2.9% in 2015, 4.7% vs. 2.0% in 2016, and 7.5% vs. 2.6% in 2017). The majority of refusals concerned both vaccinations (67.3% in 2015, 74.8% in 2016, and 68% in 2017). Among parents who refused only one vaccination, TB vaccination was refused more often than HBV (9.2% vs. 7.1% in 2015, 8.3% vs. 5.7% in 2016, and 5.9% vs. 2.7% in 2017). Similar trends were observed among the hesitants. In conclusion, it seems essential to implement effective educational and informative activities targeted to parents to reinforce positive attitudes toward vaccinations.

Keywords

Hepatitis B · Immunization · Newborns · Tuberculosis · Vaccination coverage · Vaccine

A. Nitsch-Osuch (✉)
Department of Social Medicine and Public Health, Warsaw Medical University, Warsaw, Poland

Saint Family Hospital, Warsaw, Poland
e-mail: anitsch@wum.edu.pl

B. Pawlus
Saint Family Hospital, Warsaw, Poland

M. Pawlak
Department of Social Medicine and Public Health, Warsaw Medical University, Warsaw, Poland

Voivodeship Sanitary Inspection Station, Warsaw, Poland

E. Kuchar
Department of Pediatrics and Clinical Assessment Unit, Warsaw Medical University, Warsaw, Poland

1 Introduction

According to the guidelines of the Polish National Immunization Program (PNIP), all newborns should be vaccinated against hepatitis B (HBV) and tuberculosis (TB) within the first 24 h after birth (Table 1) (PNIP 2017). These recommendations result from both previous and current epidemiological situation of infectious diseases in Poland. In the 1980s, Poland was in a group of countries with a high HBV incidence (40/100,000); however, a significant improvement in the sanitary conditions of medical facilities and the introduction of obligatory immunizations for newborns and infants has led to a revision of the epidemiological situation (Stępień et al. 2015; Magdzik and Czarkowski 2006). Likewise, a decreasing trend for TB incidence has been observed in recent years, with less than 1% prevalence in children, which shows a marginal transmission of tubercle bacilli in the general population. Nevertheless, we emphasize that the 16.7/100.000 incidence of TB in Poland in 2015 is higher than that in other EU countries (Korzeniewska-Koseła 2017). Poland is one of a few EU countries where vaccinations are administered to newborns within the first 24 h after birth. In other countries, HBV and TB vaccinations are performed at a different time or they do not belong to the child vaccination program (ECDC 2018).

Vaccinations at birth are regarded as controversial by some parents, which results in a hesitant behavior toward immunizations, a trend on a rise of late in the developed countries (Petrelli et al. 2018). Although the immunization program in Poland as yet performs at a high 95% level and involves free-of-charge vaccinations, the number of parents who avoid immunizations has sharply increased in recent years (Table 2). Coverage rates for vaccinations against HBV and TB in children aged 0–1 year throughout Poland and in the Masovian voivodship are presented in Tables 3 and 4, respectively (NIPH 2015, 2016, 2017).

The law regulates the system of reporting vaccinations in Poland. Entities responsible for immunization, including neonatal wards and primary healthcare centers, are obliged law to report to the State Sanitary Inspectorate the number and type of performed vaccinations on a quarterly basis. The aggregate data on the coverage rate for immunizations are published every year by the National Institute of Public Health – National Institute of Hygiene (NIPH-NIH) in Warsaw, Poland. These data show the percentage of vaccinated children in different age groups.

Table 1 Recommended tome of vaccination against tuberculosis and hepatitis B

Age of children		Tuberculosis	Hepatitis B
1st year of life	24 h	Yes	Yes
	2 months		Yes
	3–4 months		
	5–6 months		
	7 months		Yes
2nd year of life	13–14 months		
	16–18 months		

Table 2 Number of refusals to perform immunizations in Poland in recent years, according to the Polish National Immunization Program (PNIP 2017)

Year	Total number of refusals	Number of refusals per 1000 persons aged 0–19 years
2017	30,089	4.3
2016	23,147	3.2
2015	16,689	2.3
2014	12,681	1.7
2013	7248	0.97
2012	5340	0.70

Table 3 Hepatitis B vaccination coverage in neonates aged 0–12 months in Poland and Masovian Province in 2015–2017, according to the 0-1-6-month paradigm of vaccine administration

		First two doses[a]		Third dose[b]		Total	
Year	Infants (n)	Poland	Masovia	Poland	Masovia	Poland	Masovia
2017	5440	46.7%	47.4%	39.3%	38.4%	86.0%	85.8%
2016	5154	44.2%	44.7%	42.8%	41.8%	87.0%	86.5%
2015	4191	46.6%	48.3%	42.9%	40.1%	89.1%	88.5%

Data are percentages of neonates. [a]First dose within 24 h after birth and second dose 4 weeks after the first one; [b]Third dose at the age of 6 months

Table 4 Tuberculosis vaccination coverage in neonates aged 0–12 months in Poland and Masovian Province in 2015–2017

	Within 24 h after birth		Within 24 h–14 days of life		Within 14 days–11 months of life		In the 12th month of life		Total – in the 1st year of life	
Year	Poland	Masovia	Poland	Masovia	Poland	Masovia	Poland	Masovia	Poland	Masovia
2017	77.7%	86.2%	12.7%	5.8%	1.4%	0.8%	0.1%	0	91.8%	92.8%
2016	80.0%	87.3%	11.2%	5.2%	1.4%	0.9%	0.1%	0	92.6%	93.4%
2015	82.6%	89.6%	9.8%	4.9%	1.3%	0.9%	0.1%	0	93.8%	95.5%

Data are percentages of neonates vaccinated

However, there is no detailed information on the reasons for noncompliance with the immunization program, other than a summary information on the number of parents refusing or delaying the vaccination of children. Therefore, this study seeks to determine the timeline and completeness of vaccinations against HBV and TB in the neonates born in 2015–2017 in a single maternity hospital in Warsaw and to compare the hospital's vaccination coverage with those in the voivodship region and in the whole country.

2 Methods and Results

In this study, we retrospectively reviewed medical files of 14,785 neonates born in St. Family Hospital in Warsaw, Poland, from January 2015 to December 2017. The review was under the angle of vaccinations against HBV and TB in the first 24 h after birth. We calculated the percentage of newborns who were not vaccinated on time and those who had delayed vaccinations due to parental refusal or medical contraindications. We also compared the hospital data with vaccination coverage in the Masovian Province and with the whole of Poland.

In the 14,785 cohort of neonates, we identified 1,197 (8.1%) who were not vaccinated in the first 24 h of life according to recommended schedule. Eight hundred thirty-four out of the one thousand one hundred ninety-seven (69.7%) were not vaccinated since their parents reneged. The remaining minority of 363 (30.3%) children were not vaccinated against TB due to medical contraindication, but vaccination eventually took place before discharge from the hospital. The percentage of newborns not vaccinated against both hepatitis B and TB according to recommended schedule defined in the immunization program increased from 7.3% in 2015 to 10.1% in 2017. The increase was driven by the parental refusal to vaccinate, which increased from 4.4% in 2015 to 7.5% in 2017 year, as the proportion of newborns not vaccinated on time due to temporary medical contraindications remained at a comparable level (2.9% in 2015 vs. 2.6% in 2017) (Table 5). The majority of parents (76–84%) who failed to provide consent to vaccinate the newborn in the first 24 h after birth reaffirmed that decision later on. As a result, their children left the hospital unvaccinated. The percentage of unvaccinated newborns whose parents gave their consent for vaccination in the following days, before the discharge date,

Table 5 Hepatitis B and tuberculosis vaccinations coverage in neonates in a single hospital in 2015–2017

		Untimely vaccinations		Unvaccinated against both HVB and TB		Unvaccinated against TB		Unvaccinated against HVB	
Year	Neonates	PD	MC	PD	MC	PD	MC	PD	MC
2017	5,440 (100%)	547 (10.1%)		362 (6.7%)		169 (3.1%)		16 (0.3%)	
		408 (7.5%)	139 (2.6%)	360 (6.6%)	2 (0.1%)	32 (0.6%)	137 (2.5%)	16 (0.3%)	0
2016	5,154 (100%)	345 (6.7%)		200 (3.9%)		128 (2.5%)		17 (0.3%)	
		242 (4.7%)	103 (2.0%)	200 (3.9%)	0	25 (0.5%)	103 (1.9%)	17 (0.3%)	0
2015	4,191 (100%)	305 (7.3%)		148 (3.5%)		141 (3.4%)		16 (0.4%)	
		184 (4.4%)	121 (2.9%)	148 (3.5%)	0	20 (0.5%)	121 (2.9%)	16 (0.4%)	0

Data are counts and percentages, *n* (%). *PD* parental decision; *MC* medical contraindications

Table 6 Nonperformance of timely vaccinations of neonates against tuberculosis (TB) and hepatitis B virus (HVB) due to parents' decision in 2015–2017

		Refusal to vaccinate			Delay in vaccination		
Year	Unvaccinated neonates (n)	Both vaccines	TB	HVB	Both vaccines	TB	HVB
2017	408	312 (76%)			96 (24%)		
		277 (68.0%)	24 (5.9%)	11 (2.7%)	83 (20.4%)	8 (2.0%)	5 (1.0%)
2016	242	215 (89%)			27 (11%)		
		181 (74.8%)	20 (8.3%)	14 (5.7%)	19 (7.8%)	5 (2.0%)	3 (1.2%)
2015	184	154 (84%)			30 (16%)		
		124 (67.3%)	17 (9.2%)	13 (7.1%)	24 (13.1%)	3 (1.7%)	3 (1.7%)

amounted to 11–24% (Table 6). In the majority of cases, the decision to delay or reject a vaccination concerned both types of vaccines.

When temporary medical contraindications occurred, vaccinations were delayed and performed later than recommended in the immunization program, in line with the current practice. It is worthwhile to note that all children with the immunization postponed due to medical reasons were vaccinated at a later time, either while still in the ward or in the hospital neonatal clinic after discharge. A vast majority of delays in TB vaccination due to medical reasons stemmed from premature birth and low birth weight. There were only two newborns with the TB vaccination postponed due to mother's unclear serological status in HIV screening tests.

Proportions of newborns in a small hospital population, who were not immunized against TB according to the immunization schedule, were akin to those in the whole country, whereas the proportions of those not immunized against HVB were about two–threefold smaller than those in the whole country throughout 2015–2017 (Table 7). Proportions of nonimmunized neonates in Masovian province grossly corresponded to those in the whole country in the same period.

3 Discussion

This study shows that the number of newborns who are not administered HBV and TB vaccinations within the time frame indicated in the immunization program, due to their parents' decision, is on the rise in Poland. This phenomenon can be also observed in other countries, where the issue of reluctant or hesitant behavior toward immunization is growing. In most regions

Table 7 Comparison of the proportion of neonates unvaccinated against hepatitis B virus (HVB) and tuberculosis (TB) in the population of a single hospital, the Masovian province, and the whole country in 2015–2017

Year	Against TB			Against HVB		
	Hospital	Masovia	Poland	Hospital	Masovia	Poland
2017	9.8%	7.2%	8.2%	7.0%	14.2%	14%
2016	6.3%	6.6%	7.4%	4.2%	13.5%	13%
2015	6.9%	4.5%	6.2%	3.9%	11.5%	10.9%

of the US, proportion of children aged under 35 months with a complete primary series of vaccination is below the Healthy People 2020 goal (CDC 2013). In Italy, 83.7% of parents are pro-vaccine, 15.6% are vaccine-hesitant, and 0.7% are anti-vaccine (Giambi et al. 2018).

Leask et al. (2012) have identified five distinct parental groups in developed countries: "unquestioned acceptors" (30–40%), "cautious acceptors" (25–35%), "hesitant" (2–27%), "late or selective vaccinators", and refusers (<2%). In the present study, parents who refused immunization of newborns were regarded as "refusers" when the neonate left the hospital unvaccinated and as "late or selective vaccinators" when the parents consented to immunization against one or both diseases before discharge but the moment of vaccination failed to follow the obligatory standard of the immunization program and it was kept discretionary. It cannot be excluded that some parents, who had refused vaccinations during the hospital stay, decided to vaccinate their children after leaving the neonatal ward, within the framework of primary healthcare, thus becoming "late or selective vaccinators." A lack of information on the implementation of the immunization program after discharge from the neonatal ward is a limiting factor for a full interpretation of this study.

It is hard to compare the present results with data from other European countries, as universal vaccinations against HVB and TB in the first 24 h after birth are recommended in just a few countries such as Poland, Lithuania, Portugal, and Romania. In other countries, HVB vaccination is done in the risk groups only or in the neonates whose mothers are infected with hepatitis B virus (ECDC 2018). Harmsen et al. (2012) have shown that the majority of parents in the Netherlands approve of HVB vaccinations within 24 h after birth, introduced as of 2001. Likewise, universal TB vaccination in the first days after birth is recommended in Romania (2–7 day), Slovenia, Estonia (2–5 day), Croatia, Bulgaria, Ireland, Lithuania, Latvia, and Hungary (ECDC 2018). In other countries, vaccination against TB is not obligatory or it is performed only in risk groups. According to Chiabi et al. (2017), timely TB vaccination in the newborn in Cameroon is done in 83% of children, while delays and refusals result from the parental decisions (mainly father's) and from the mother's level of education. Parache et al. (2010) have concluded that the number of parental refusals for TB vaccination in Italy is lower in neonates belonging to high risk groups. A Dutch study has shown that the proportion of parents who refuse vaccinations amounts to 9% (Hontelez et al. 2010), which is akin to the 2017 figure presented in the present study. A Polish cross-sectional study of Braczkowska et al. (2018) has pointed out that parents who refuse to accept the safety of immunization opine that vaccination should not be performed too early and the number of vaccinations should be reduced.

The results of this study indicate that medical reasons for postponing vaccinations within 24 h after birth are of temporary character in the majority of cases as these vaccinations are administered at a later time. Furthermore, proportion of children not vaccinated on time due to medical contraindications remains at a stably low level, which points to a good classification of newborns for vaccination and shows that contraindications are not overused as a reason to delay immunization at birth. There is a substantial improvement in the vaccination process from the medical standpoint that has recently taken root in Poland as

Mrożek-Budzyn (2001) reported about two decades ago that it was doctors at the time who led to delays in vaccinations about twice more frequent than parents, most often mothers, did.

A limitation of our study is a lack of information on vaccinations performed after discharge of neonates from the hospital. Another limitation stems from differences in the way vaccinations against HVB and TB are reported at the regional and country levels. There are no general data on the percentage of newborns unvaccinated against HVB. The available data concern the primary course of immunization which includes the first two doses of vaccine given at an interval of at least 4 weeks. In case of TB, data on vaccination coverage at birth are available for specific age groups such as 0–14 days and 0–12 months of life. However, the reporting system does not allow for the identification of reasons for a delay in vaccination.

We conclude that the increasing number of newborns who are not vaccinated at birth against hepatitis B and tuberculosis, due to their parents' decision, is highly disturbing as it is incompatible with contemporary medical knowledge and healthcare rules as well as with the National Immunization Program in Poland. Therefore, it is essential to intensify educational activity addressed to future potential parents and expectant parents on the biomedical meaning and value of immunization in order to dispel doubts of hesitant parents and to counter the anti-vaccination attitude.

Conflicts of Interest The authors declare no conflict of interests in relation to this article.

Ethical Approval All procedures performed in studies involving human participants were in accordance with the ethical standards of the institutional and/or national research committee and with the 1964 Helsinki declaration and its later amendments or comparable ethical standards. The study was approved by the Ethics Committee of Warsaw Medical University in Poland.

Informed Consent The requirement to obtain individual patient consent was waived as the study is of retrospective nature consisting of the anonymous review of patient hospital files.

References

Braczkowska B, Kowalska M, Barański K, Gajda M, Kurowski T, Zejda JE (2018) Parental opinions and attitudes about children's vaccination safety in Silesian Voivodeship, Poland. Int J Environ Res Public Health 15:224–228

CDC (2013) Centers for Disease Control and Prevention. National, state, and local area vaccination coverage among children aged 19–35 months – United States, 2012. MMWR Morb Mortal Wkly Rep 62 (36):733–740

Chiabi A, Nguefack FD, Njapndounke F, Kobela M, Kenfack K, Nguefack S, Mah E, Nguefack-Tsague G, Angwafo F (2017) Vaccination of infants aged 0 to 11 months at the Yaounde Gynaeco-obstetric and pediatric hospital in Cameroon: how complete and how timely? BMC Pediatr 17:206

ECDC (2018) European Center for Diseases Control and Prevention. Vaccine scheduler. https://vaccine-scheduler.ecdc.europa.eu/Scheduler/ByDisease?SelectedDiseaseId=6&SelectedCountryIdByDisease=-1. Accessed on 2 May 2018

Giambi C, Fabiani M, D'Ancona F, Ferrara L, Fiacchini D, Gallo T, Martinelli D, Pascucci MG, Prato R, Filia A, Bella A, Del Manso M, Rizzo C, Rota MC (2018) Parental vaccine hesitancy in Italy – results from a national survey. Vaccine 36:779–787

Harmsen IA, Lambooij MS, Ruiter RA, Mollema L, Veldwijk J, van Weert YJ, Kok G, Paulussen TG, de Wit GA, de Melker HE (2012) Psychosocial determinants of parents' intention to vaccinate their newborn child against hepatitis B. Vaccine 30:4771–4777

Hontelez JA, Hahné SJ, Oomen P, de Melker H (2010) Parental attitude towards childhood HBV vaccination in the Netherlands. Vaccine 28:1015–1020

Korzeniewska-Koseła M (2017) Tuberculosis in Poland in 2015. Przegl Epidemiol 71:391–403

Leask J, Kinnersley P, Jackson C, Cheater F, Bedford H, Rowles G (2012) Communicating with parents about vaccination: a framework for health professionals. BMC Pediatr 12:154

Magdzik W, Czarkowski MP (2006) Epidemiological situation of hepatitis B in Poland in the years 1979–2004. Przegl Epidemiol 60:471–480

Mrożek-Budzyn D (2001) The reasons of low performance and delaying in the compulsory vaccinations in children. Przegl Epidemiol 55:343–353

NIPH-NIH (2015) National Institute of Public Health – National Institute of Hygiene Vaccinations in Poland in 2015. https://wwwold.pzh.gov.pl/oldpage/epimeld/2015/Sz_2015.pdf. Accessed on 14 Mar 2018

NIPH-NIH (2016) National Institute of Public Health - National Institute of Hygiene Vaccinations in Poland in 2016. https://wwwold.pzh.gov.pl/oldpage/epimeld/2016/Sz_2016.pdf. Accessed on 14 Mar 2018

NIPH-NIH (2017) National Institute of Public Health – National Institute of Hygiene Vaccinations in Poland in 2017. https://wwwold.pzh.gov.pl/oldpage/epimeld/2017/Sz_2017_wstepne_dane.pdf. Accessed on 22 Oct 2018

Parache C, Carcopino X, Gossot S, Retornaz K, Uters M, Mancini J, Garnier JM, Minodier P (2010) Bacillus Calmette-Guérin (BCG) vaccine coverage in newborns and infants at risk before and after a change in BCG policy. Arch Pediatr 17:359–365

Petrelli F, Contratti CM, Tanzi E, Grappasonni I (2018) Vaccine hesitancy, a public health problem. Ann Ig 30:86–103

PNIP (2017) Polish national immunization program. https://gis.gov.pl/images/ep/so/pso_2017-_nowelizacja.pdf. Accessed on 14 Mar 2018

Stępień M, Piwowarow K, Czarkowski MP (2015) Hepatitis B in Poland in 2015. Przegl Epidemiol 71:351–362

Adv Exp Med Biol - Clinical and Experimental Biomedicine (2020) 8: 107–113
https://doi.org/10.1007/5584_2019_454

Published online: 13 December 2019

Virological and Epidemiological Situation in the Influenza Epidemic Seasons 2016/2017 and 2017/2018 in Poland

E. Hallmann-Szelińska, K. Łuniewska, K. Szymański, D. Kowalczyk, R. Sałamatin, A. Masny, and L. B. Brydak

Abstract

The World Health Organization estimates that influenza virus infects 3–5 million people worldwide every year, of whom 290,000 to 650,000 die. In the 2016/2017 epidemic season in Poland, the incidence of influenza was 1,692 per 100,000 population. The influenza A virus, subtype A/H3N2/, was the predominant one in that season. However, in the most recent 2017/2018 epidemic season, the incidence exceeded 1,782 per 100,000 already by August of 2018. In this season, influenza B virus predominated, while the A/H1N1/pdm09 strain was most frequent among the influenza A subtypes. The peak incidence, based on the number of clinical specimens tested, was in weeks 4–5 of 2017 and week 8 of 2018 in the 2016/2017 and 2017/2018 epidemic seasons, respectively. As of the 2017/2018 season, a quadrivalent vaccine, consisting of two antigens of influenza A subtypes and another two of influenza B virus, was available in Poland. Nonetheless, the vaccination rate remained at one of the lowest level in Europe, fluctuating between 3% and 4% of the general Polish population.

Keywords

Disease incidence · Epidemic season · Epidemiology · Infection · Influenza · Vaccination rate · Viral subtypes

1 Introduction

The influenza virus belongs to the *Orthomyxoviridae* family. It infects epithelial cells of the nose, larynx, trachea, and bronchi, damaging the epithelium of the respiratory system. The influenza A virus consists of eight RNA segments coding for at least ten proteins (Brydak 2008). It is divided into subtypes based on the properties of surface antigens, i.e., neuraminidase (NA) and hemagglutinin (HA). There are 18 antigenic HA subtypes (H1–H18) and 11 NA subtypes (N1–N11) (Tong et al. 2013). Type B influenza virus is not categorized into subtypes. It is divided into two phylogenetically and antigenically different lines: Victoria and Yamagata, which are circulating worldwide in various proportions (Webster et al. 2013).

The epidemic season lasts from the beginning of October of a year to the end of September the

E. Hallmann-Szelińska (✉), K. Łuniewska, K. Szymański, D. Kowalczyk, A. Masny, and L. B. Brydak
Department of Influenza Research – National Influenza Center, National Institute of Public Health – National Institute of Hygiene, Warsaw, Poland
e-mail: ehallmann@pzh.gov.pl

R. Sałamatin
Department of General Biology and Parasitology, Warsaw Medical University, Warsaw, Poland

following year (52 weeks). The annual influenza epidemic affects 3–5 million people worldwide, 20–30% of whom are children. Influenza is a health scourge causing severe complications. There are around 290,000–650,000 deaths yearly. Influenza B virus most often affects children and young adolescents (0–17 years of age) and adults of 25–44 years of age (Vijaykrishna et al. 2015; Socan et al. 2014). Unlike the influenza A virus, the influenza B virus does not cause pandemics (Caini et al. 2015).

Poland has been a member of the European Influenza Surveillance Network (EISN) since 2001. Since the 2004/2005 epidemic season, it has also participated in the Sentinel surveillance system. Virological and epidemiological data had initially been recorded in four age groups: 0–4, 5–14, 15–64, and $\geq$ 65 years of age. The system has been revised as of the 2013/2014 season to narrow the age group range to 0–4, 5–9, 10–14, 15–25, 26–44, 45–64, and $\geq$ 65 years of age (Bednarska et al. 2016) in order to better illustrate the epidemiological situation.

The aim of this study was to investigate the virological and epidemiological situation during the two successive influenza epidemic seasons in Poland of 2016/2017 and 2017/2018.

2 Methods

The evaluation of the epidemiological situation in the epidemic seasons 2016/2017 and 2017/2018 was based on the data obtained from weekly epidemiological reports prepared jointly by the Department of Influenza Virus Research, National Center for Influenza of the National Institute of Public Health-National Institute of Hygiene (NIPH-NIH), and the Chief Sanitary Inspectorate in Warsaw, Poland. Reports concerned the incidence of both clinically and laboratory-confirmed influenza and influenza-like illnesses and also acute respiratory tract infections, according to the criteria set for the surveillance of influenza in the EU. Virological diagnostic tests were performed in the laboratories of the Provincial Sanitary and Epidemiological Stations and were verified in the NIPH-NIH within the influenza surveillance Sentinel and non-Sentinel systems. The tests were performed on clinical samples collected by primary care physicians from patients suspected of being infected with influenza and influenza-like viruses. Molecular methods were employed to define the genetic viral material.

The genetic material was isolated from 200 μL pharyngeal swab specimens suspended in physiological saline or phosphate-buffered saline (PBS). RNA isolations were performed using a Maxwell 16 Viral Total Nucleic Acid Purification Kit (Promega Corporation, Madison, WI). The procedure was carried out in accordance with the manufacturer's instructions. Molecular analysis was performed in a LightCycler 2.0 instrument (Roche Diagnostics, Rotkreuz, Switzerland). The reaction mixture contained the following ingredients in 20 μL capillaries: $MgSO_4$, bovine serum albumin (BSA), reaction buffer, RNAse-free water, SuperScript®III/Platinum®Taq Mix (Invitrogen Life Technologies – Thermo Fisher Scientific, Carlsbad, CA), 0.5 μL probe, and 0.5 μL (20 nM) primers. Probes and primers were obtained from the International Reagent Resource (IRR) established by the Centers for Disease Control and Prevention (CDC). Five microliters of RNA was added to each sample of reaction mixture. Negative control was RNAse-free water, while positive control was RNA of strains included in the influenza vaccine for a given epidemic season.

To determine the influenza-like virus infection, clinical material was investigated using an RV15 OneStep ACE Detection Kit (Seegene, Seoul, South Korea). This multiplex kit detects genetic material of the following respiratory viruses: influenza types A and B; adenovirus (ADV); respiratory syncytial viruses (RSV) A and B; human metapneumovirus (hMPV); human coronavirus (hCoV); human parainfluenza viruse (hPIV) types: 1, 2, 3, and 4; human bocavirus (hBoV); and enterovirus (EV).

Data were expressed as counts of the incidence per 100,000 population in a given epidemic season, as total, and by age groups. Differences between the percentages of confirmed cases of influenza infection between the corresponding age groups of the two epidemic seasons were assessed with a two-sample *t*-test. A p-value of <0.05 defined statistically significant differences.

3 Results and Discussion

In the 2016/2017 epidemic season, i.e., from 01/10/2016 to 30/09/2017, a total of 3,977 specimens were tested in Poland, with a positive diagnostic yield of 44.9%. Among positive samples, influenza viruses accounted for 94.5%, whereas influenza-like viruses accounted for 5.5% of the infections. Virological molecular analysis revealed that 97.6% of influenza infections in the whole population were caused by type A virus, of which 66.8% were unsubtyped A virus and 31.2% were subtype A/H3N2/. The remaining 2% were infections caused by influenza type B virus. RSV predominated among influenza-like viruses (85.7%). Figure 1 presents the percentage distribution of influenza viruses by age groups in the 2017/2018 season. The highest percentage of confirmed cases of influenza A infections was in the age group of 26–44 years (99.1%), while that for influenza type B was in the age group of 5–9 years (6.8%).

For comparison, in the 2017/2018 epidemic season, 5,793 specimens were tested, with a positive diagnostic yield of 43.8%. In contrast to the preceding season when type A virus predominated, in this season type B virus was a dominant contagion. It accounted for 33.5% of confirmed influenza infections, whereas type B virus accounted for 63.3% and influenza-like viruses for 3.2% infections in the whole population. Among the subtypes of influenza A virus, subtype A/H1N1/pdm09 predominated in 85.9% of cases. RSV predominated in 63.1% of influenza-like infections. Figure 2 presents the percentage distribution of influenza viruses by age groups in the 2017/2018 season. The highest percentage of confirmed cases of influenza A infections was in the age group of 0–4 years (66.8%), while that for influenza type B was in the age group of 65+ years (76.6%).

The number of confirmed cases of influenza infections, relative to the number of samples tested, was significantly larger in the 2017/2018 season, compared with the preceding season, in the 0–4 years ($p < 0.001$), 5–9 years ($p = 0.003$), and 65+ years ($p = 0.001$) of age groups.

In the 2016/2017 season, overall 12 coinfection were detected: 9 between influenza viruses,

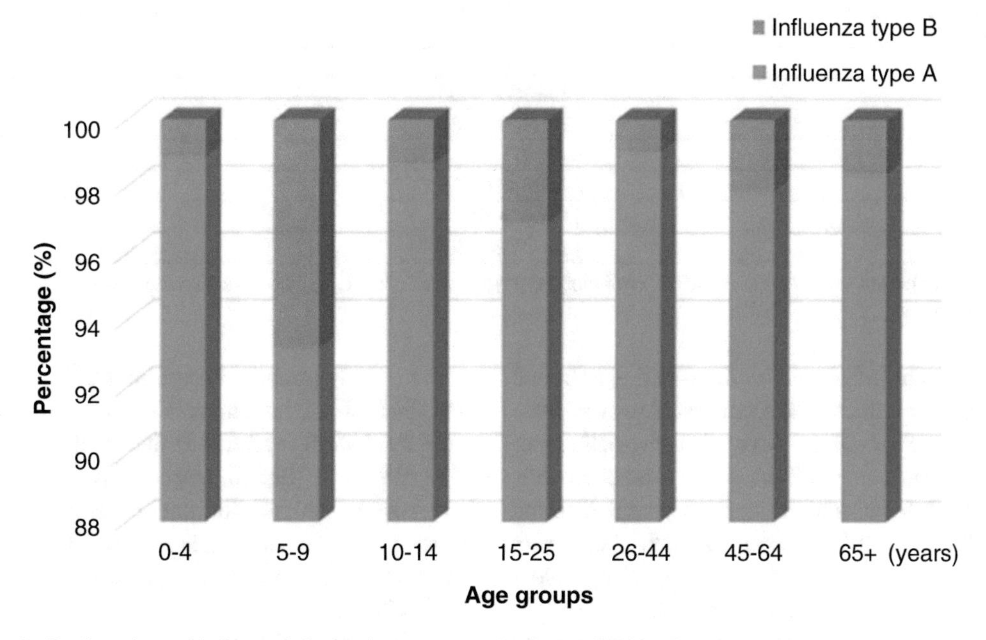

Fig. 1 Confirmations of influenza infection by age groups in the 2016/2017 epidemic season

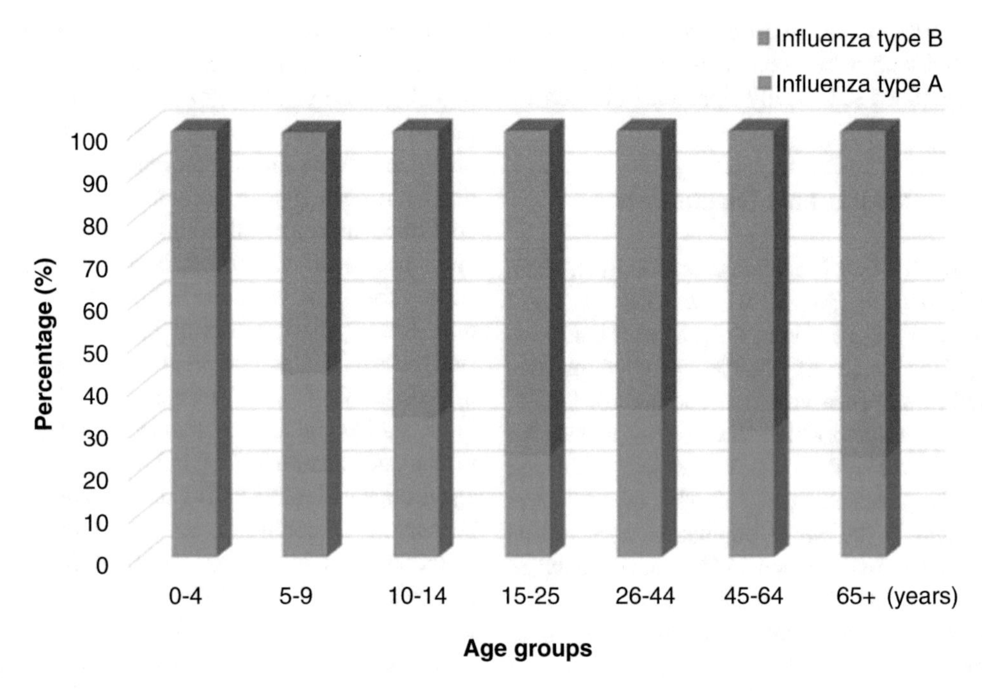

Fig. 2 Confirmations of influenza infection by age groups in the 2017/2018 epidemic season

Table 1 Coinfections in the 2016/2017 epidemic season

Patient age (years)	Type of coinfection	Epidemic week
56	A/H3N2/ + B	50
70	A/H3N2/ + B	50
4	A/H3N2/ + B	51
70	A/H3N2/ + B	51
50	A/H3N2/ + B	51
67	A/H3N2/ + B	51
8	A/H3N2/ + B	1
79	A/H3N2/ + B	2
4	A + B	3
70	A/H3N2/ + RSV	4
63	PIV-1 + EV	5
4	A/H3N2/ + RSV	7

A + B, influenza virus A and B; RSV, respiratory syncytial virus; PIV-1, human parainfluenza virus type 1; EV, enteroviruses

2 between influenza A virus and RSV, and 1 between influenza-like viruses. However, there were no coinfections detected in the patients of 26–44 years of age (Table 1). On the other side, three coinfections in this age group were detected in the 2017/2018 season. Overall, there were 17 coinfections detected in the later season: 11 between influenza viruses, 2 between influenza viruses and influenza-like viruses, and 4 between influenza-like viruses (Table 2).

The overall number of suspicions of influenza and influenza-like infections in Poland was larger in the 2017/2018 season than that in 2016/2017: 5,337,619 vs. 4,919,110, respectively. Likewise, there were 18,320 hospitalizations and 48 deaths recorded due to influenza in the 2017/2018 season vs. 16,890 hospitalizations and 25 deaths in 2016/2017.

In the 2016/2017 season, the incidence of influenza was 1,692 per 100,000 population.

Table 2 Coinfections in the 2017/2018 epidemic season

Patient age (years)	Type of coinfection	Epidemic week
16	RSV + hCoV	2
7	A/H3N2/ + B	3
9	A/H3N2/ + B	4
70	A/H3N2/ + B	5
5	hCoV + PIV 1, 2, 3	5
5	PIV-1 + PIV 2	7
64	A/H1N1/pdm09 + B	8
33	RSV + B	8
4	A/H3N2/ + B	9
2	RSV + B	10
3	A/H1N1/pdm09 + B	10
71	A + B	10
46	A + B	10
3	PIV- 1 + hMPV	10
40	A + B	11
35	A/H3N2/ + B	11
69	A + B	12

RSV, respiratory syncytial virus; hCoV, human coronavirus; A + B, influenza virus A and B; PIV 1, 2, and 3, human parainfluenza virus type 1, 2, and 3; hMPV, human metapneumovirus

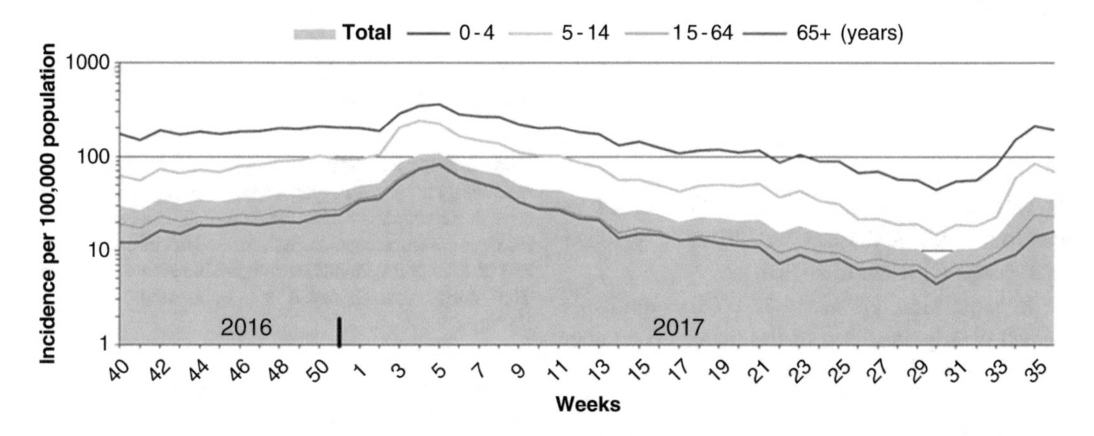

Fig. 3 Influenza and influenza-like infections in the 2016/2017 epidemic seasons in the Polish population by age groups, according to the National Institute of Public Health-National Institute of Hygiene in Warsaw

The influenza A virus, subtype A/H3N2/, was the predominant one in that season. However, in the most recent 2017/2018 epidemic season, the incidence was greater, exceeding 1,782 per 100,000 already by August of 2018. In this season, influenza B virus predominated, while A/H1N1/ pdm09 strain was most frequent among the influenza A subtypes. The peak incidence, based on the number of clinical specimens tested, was in weeks 4–5 of 2017 and week 8 of 2018 in the respective seasons (Figs. 3 and 4).

Across the whole of Europe, virus type A subtype A/H3N2/predominated in the 2016/2017 season (ECDC 2017), which was akin to the epidemiological situation at the time in Poland. However, in the following 2017/2018 season, the predominance switched to type B Yamagata lineage virus in Europe. Among somehow less

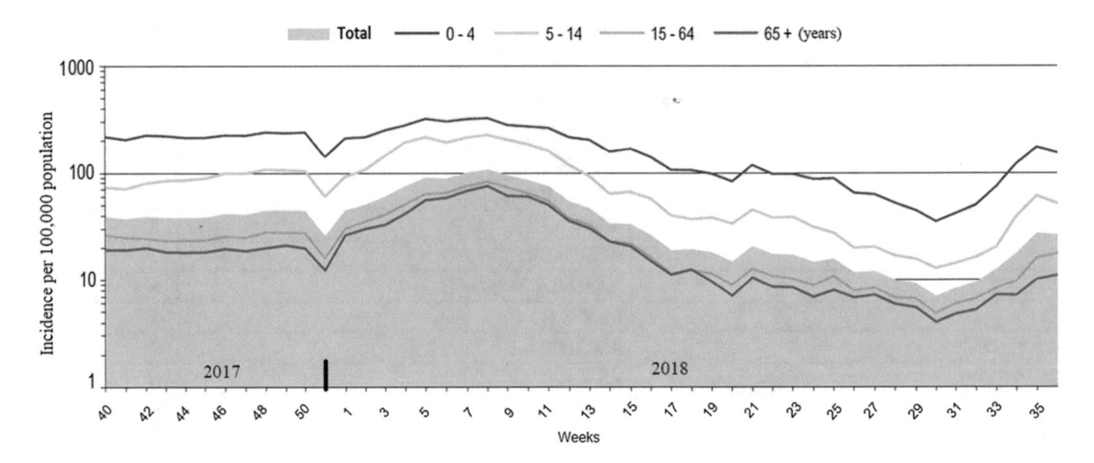

Fig. 4 Influenza and influenza-like infections in the 2017/2018 epidemic seasons in the Polish population by age groups, according to the National Institute of Public Health-National Institute of Hygiene in Warsaw

frequent influenza A infections, confirmed in 37% of all cases, circulation of A/H1N1/pdm09 and A/H3N2/strains was most commonly noticed (Adlhoch et al. 2018; ECDC 2018). In this season, the epidemic was longer than usual (Groeneveld et al. 2018). Influenza A/H3N2/virus infections are considered more severe and more often affecting people over the age of 65, leading to outbreaks in long-term nursing homes and causing a high rate of hospitalization and mortality in this age group. B viruses, on the other side, are described as causing a milder disease and affecting younger age groups (Webster et al. 2013).

In conclusion, the recent 2017/2018 epidemic season clearly consisted of the increased number and severity of influenza infections, with a shift from type A to type B virus predominance. This report underscores a fleeting viral multifariousness of influenza infection and the need to prepare adequate anti-influenza shots every season. To this end, a quadrivalent vaccine, consisting of two antigens of influenza A subtypes and another two of type B virus, was introduced in Poland as of the 2017/2018 season. Nonetheless, the vaccination rate remained at a dismal level, one of the lowest in Europe, fluctuating between 3% and 4% of the general Polish population. Thus, the epidemiological situation calls for the continuing promotion of awareness and education in the fight against influenza infection through the vaccination programs across all of the population segments.

Acknowledgments Funded by NIPH-NIH thematic subject 4/EM. We would like to acknowledge physicians and employees of the Voivodship Sanitary Epidemiological Stations across the country who collected epidemiological data for the Sentinel program in the framework of influenza surveillance in Poland.

Conflicts of Interest The authors declare no conflicts of interest in relation to this article.

Ethical Approval All procedures performed in studies involving human participants were in accordance with the ethical standards of the institutional and/or national research committee and with the 1964 Helsinki declaration and its later amendments or comparable ethical standards. The study was approved by an institutional Ethics Committee.

Informed Consent Informed consent was obtained from all individual participants included in the study before collection of nasopharyngeal samples.

References

Adlhoch C, Snacken R, Melidou A, Ionescu S, Penttinen P, The European Influenza Surveillance Network (2018) Dominant influenza A(H3N2) and B/Yamagata virus circulation in EU/EEA, 2016/17 and 2017/18 seasons, respectively. Euro Surveill 23(13)

Bednarska K, Hallmann-Szelińks E, Kondratiuk K, Rabczenko D, Brydak LB (2016) Novelties in influenza surveillance in Poland. Probl Hig Epidemiol 97(2):101–105

Brydak LB (2008) Influenza, pandemic flu, myth or real threat? Rythm, Warsaw, pp 1–492. (Book in Polish)

Caini S, Huang QS, Ciblak MA, Kusznierz G, Owen R, Wangchuk S, Henriques CM, Njouom R, Fasce RA, Yu H, Feng L, Zambon M, Clara AW, Kosasih H, Puzelli S, Kadjo HA, Emukule G, Heraud JM, Ang LW, Venter M, Mironenko A, Brammer L, Mai le TQ, Schellevis F, Plotkin S, Paget J, Global Influenza B Study (2015) Epidemiological and virological characteristics of influenza B: results of the global influenza B study. Influenza Other Respir Viruses 1 (9 Suppl):3–12

ECDC (2017) European center for disease prevention and control. Influenza in Europe, summary of the season 2016–2017. https://ecdc.europa.eu/en/seasonal-influenza/season-2016–17. Accessed on 24 June 2019

ECDC (2018) European center for disease prevention and control. Influenza in Europe, summary of the season 2017–2018. https://ecdc.europa.eu/en/seasonal-influenza/season-2017–18. Accessed on 24 June 2019

Groeneveld GH, Spaan WJ, van der Hoek W, van Dissel JT (2018) The severe flu season of 2017–2018: making a case for the vaccination of healthcare professionals. Ned Tijdschr Geneeskd 6:162

Socan M, Prosenc K, Ucakar V, Berginc N (2014) A comparison of the demographic and clinical characteristics of laboratory confirmed influenza B Yamagata and Victoria lineage infection. J Clin Virol 61(1):156–156

Tong S, Zhu X, Lil Y, Shi M, Zhang J, Bourgeois M, Yang H, Chen X, Recuenco S, Gomez J, Chen LM, Johnson A, Tao Y, Dreyfus C, Yu W, McBride R, Carney PJ, Gilbert AT, Chang J, Guo Z, Davis CT, Paulson JC, Stevens J, Rupprecht CE, Holmes EC, Wilson IA, Donis RO (2013) New world bats harbor diverse influenza a viruses. PLoS Pathog 9(10): e1003657

Vijaykrishna D, Holmes EC, Joseph U, Fourment M, Su YC, Halpin R, Lee RT, Deng YM, Gunalan V, Lin X, Stockwell TB, Fedorova NB, Zhou B, Spirason N, Kühnert D, Bošková V, Stadler T, Costa AM, Dwyer DE, Huang QS, Jennings LC, Rawlinson W, Sullivan SG, Hurt AC, Maurer-Stroh S, Wentworth DE, Smith GJ, Barr IG (2015) The contrasting phylodynamics of human influenza B viruses. elife 4:e05055

Webster RG, Monto AS, Braciale TJ, Lamb RA (2013) Textbook of influenza, 2nd edn. Wiley, Hoboken, pp 392–418

Adv Exp Med Biol - Clinical and Experimental Biomedicine (2020) 8: 115–121
https://doi.org/10.1007/5584_2019_462

Published online: 28 January 2020

Epidemic Influenza Seasons from 2008 to 2018 in Poland: A Focused Review of Virological Characteristics

Sainjargal Byambasuren, Iwona Paradowska-Stankiewicz, and Lidia B. Brydak

Abstract

The objective of this review was to elaborate on changes in the virological characteristics of influenza seasons in Poland in the past decade. The elaboration was based on the international influenza surveillance system consisting of Sentinel and non-Sentinel programs, recently adopted by Poland, in which professionals engaged in health care had reported tens of thousands of cases of acute upper airway infections. The reporting was followed by the provision of biological specimens collected from patients with suspected influenza and influenza-like infection, in which the causative contagion was then verified with molecular methods. The peak incidence of influenza infections has regularly been in January–March each epidemic season. The number of tested specimens ranged from 2066 to 8367 *per* season from 2008/2009 to 2017/2018. Type A virus predominated in nine out of the ten seasons and type B virus of the Yamagata lineage in the 2017/2018 season. Concerning the influenza-like infection, respiratory syncytial virus predominated in all the seasons. There was a sharp increase in the proportion of laboratory confirmations of influenza infection from season to season in relation to the number of specimens examined, from 3.2% to 42.4% over the decade. The number of confirmations, enabling a prompt commencement of antiviral treatment, related to the number of specimens collected from patients and on the virological situation in a given season. Yet influenza remains a health scourge, with a dismally low yearly vaccination rate, which recently reaches just about 3.5% of the general population in Poland.

Keywords

Diagnostics · Epidemic seasons · Influenza · Sentinel program · Surveillance · Virological characteristics

S. Byambasuren
EASTMED Family Medicine, Acupuncture, Aesthetic Medicine, Warsaw, Poland

I. Paradowska-Stankiewicz (✉)
Department of Epidemiology of Infectious Diseases and Surveillance, National Institute of Public Health – National Institute of Hygiene, Warsaw, Poland
e-mail: istankiewicz@pzh.gov.pl

L. B. Brydak
Department of Influenza Research, National Influenza Center at the National Institute of Public Health - National Institute of Hygiene in Warsaw, Warsaw, Poland

1 Introduction

Influenza is a contemporary worldwide scourge. There were three influenza pandemics in the nineteenth century. The most tragic one, the Spanish flu, caused 50–100 m deaths and considerable economic losses, according to the contemporary estimates. The WHO proposed a global program for the surveillance of influenza in 1947, based on the epidemiological and virological research (Bednarska et al. 2016a; Broberg et al. 2015). This program has currently assumed a form of the Global Influenza Surveillance and Response System, Poland has been participating in since 1957. The surveillance system has included the Sentinel program as of 2004. The surveillance is pursued in cooperation with general practitioners and laboratories of 16 Provincial Sanitary-Epidemiological Stations across the country. The epidemiological part of this surveillance is based on reports of cases and suspected cases of influenza and influenza-like illnesses. The virological monitoring is based on a laboratory confirmation of infection using molecular biology methods (Bednarska et al. 2016b). In addition, virological data recorded in the non-Sentinel program are collected from hospitals and private physicians.

Influenza infection, with the occurrence of complications, mortality, and absenteeism at work, bears outstanding health and socioeconomic burdens. This study was designed to present and compare the virological characteristics of influenza seasons in Poland in the last decade, based on the epidemiological data from the national influenza surveillance system.

2 Material and Results

This article was based on the influenza surveillance system consisting of Sentinel and non-Sentinel programs, recently adopted by Poland, in which health-care professionals are obliged to report cases of acute upper airway infections. Preliminary diagnosis was done in 16 Provincial Sanitary-Epidemiological Stations throughout the country, followed by the sampling of nasopharyngeal specimens in case of suspected influenza and influenza-like infections. The reports were collated and coordinated by the Department of Influenza Research of the National Influenza Center at the National Institute of Public Health-National Institute of Hygiene (NIPH-NIH) in Warsaw, where the causative contagion was then verified using molecular methods. Data were collected at weekly cycles for 52 weeks beginning as of October and ending in September of a successive year. Yearly number of collected and tested specimens in the surveillance system ranged from 2066 to 8367 during the 10-year period reviewed in this study (Cieślak et al. 2017, 2018).

The study material consisted of nasopharyngeal swabs or bronchial tree lavage collected from patients presenting symptoms of upper airway infection across all age groups. Specimens were collected through the epidemic seasons from 2008/2009 to 2017/2018. They were further investigated using real-time polymerase chain reaction (PCR) or conventional multiplex PCR to isolate genetic material of invading contagions and thus to identify the influenza or influenza-like type of virus.

In total, during the ten epidemic seasons, physicians collected and transferred to the NIPH-NIH center 44,566 samples, which gives an average of 4456 samples *per* season (Table 1). In the 2008/2009–2017/2018 epidemic seasons, influenza type A and B viruses and influenza-like illnesses of different severity were recorded. The percentage of confirmed influenza virus infections ranged from 21.4% in 2011/2012 to 96.7% in 2017/2018. There was an apparent inverse relation between the frequency of influenza and influenza-like infections; the more of the former, the fewer of the latter. In nine out of the ten seasons, type A virus predominated, with subtype A/H1N1/pdm09 in seven and A/H3N2/ in two seasons. The influenza type B virus of the Yamagata lineage predominated in the 2017/2018 season (Table 1).

Infections caused by influenza-like viruses were also recorded. The following viruses were identified: human respiratory syncytial virus (RSV) type A and type B, parainfluenza virus types 1–4, human metapneumovirus (hMV), human adenovirus (ADV), human rhinovirus (RV), human coronavirus 229/NL63 and OC43/HKU1, and enterovirus. The percentage of confirmed influenza-like infections among all influenza infections ranged from 78.6% in 2011/2012

Table 1 Virological characteristics of influenza and influenza-like infections during a decade of the 2008/2009–2017/2018 epidemic seasons in Poland

Epidemic season	No. of samples with confirmed infection	Percentage of influenza/ influenza-like viral infections (%)	Predominant subtype of influenza virus (%)	Predominant influenza-like virus (%)
2008/ 2009	3379	56.8/43.2	A/H1N1/pdm09 (54.5)	RSV (54.6)
2009/ 2010	6185	87.3/12.7	A/H1N1/pdm09 (73.2)	RSV (56.7)
2010/ 2011	2718	76.3/23.7	A/H1N1/pdm09 (67.1)	RSV (74.0)
2011/ 2012	2066	21.4/78.6	A/H1N1/pdm09 (24.3)	RSV (86.7)
2012/ 2013	6949	75.8/24.2	A/H1N1/pdm09 (58.5)	RSV (87.0)
2013/ 2014	2620	42.7/57.3	A/H3N2/ (30.6)	RSV (84.5)
2014/ 2015	2512	68.7/31.3	A/H1N1/pdm09 (13.2)	RSV (91.4)
2015/ 2016	8367	94.3/5.7	A/H1N1/pdm09 (49.7)	RSV (77.5)
2016/ 2017	3977	94.5/5.5	A/H3N2/ (29.7)	RSV (87.5)
2017/ 2018	5793	96.7/3.3	B lineage Yamagata (65.5)	RSV (63.1)

to 3.3% in 2017/2018. It is worthy of note that RSV infection predominated in each epidemic season, with the confirmation rate ranging from 54.6% in 2008/2009 to 91.4% in 2014/2015 among all the other influenza-like viruses. Interestingly, influenza-like infections were more frequent than true influenza infections in the 2013/2014 season (Table 1).

3 Discussion

One of the basic goals of epidemiological and virological surveillance, which has been initiated in 1947, is to protect humanity from the threat of influenza. This goal has been reiterated in the 2019 publication of WHO entitled "Global Influenza Strategy 2019–2030" (WHO 2019). The surveillance is designed to monitor the spread of influenza virus from animals to humans. Molecular biology has been used to define the type of virus that circulates in a population in a given epidemic season as a candidate for the optimum seasonal composition of influenza vaccine. Further, influenza and influenza-like infection morbidity, comorbidity, and death have been recorded and expressed as the incidence of confirmed and suspected cases, the number of referrals to hospital, and variable activities of circulating influenza A and B type and influenza-like viruses. There are over 200 respiratory influenza-related viruses that circulate in the population. Therefore, virological surveillance is essential from both confirmatory and clinical standpoints. The latter has to do with advance preparation of proper vaccine composition for each upcoming season, taking into account the *genetic* and *antigenic* diversity of recently circulating influenza viruses as well as with the use of recently available novel antiviral medicines.

In the past, Poland used to send reports to the WHO Collaborating Center for Reference and Research on Influenza of the National Institute for Medical Research in London and then as of the 2015/2016 season to the WHO Collaborating Center for Reference and Research on Influenza of the Francis Crick Institute in London. On the basis of reports submitted by national influenza centers, virological situation can be traced in a given continent and also worldwide. In its current form, the Global Influenza Surveillance and Response System consists of 6 WHO Collaborating Centers for Reference and Research and 149 National Influenza Centers, including 1 in the Department of Influenza Research of the National Influenza Center at the NIPH-NIH. Their main task is to send epidemiological and virological reports and full documentation on selected isolated influenza viruses to relevant WHO reference centers.

In the case of the 2008/2009 season, the peak of influenza in both Poland and Europe fell on the 1st week of February 2009 (MRC 2008). Influenza A/H1N1/pdm09 was the predominant type A virus, which was confirmed in Europe in 53% of the submitted reports. Likewise, the confirmation rate for this virus amounted to 54.5% in Poland. There were no influenza-related deaths recorded in Poland in that season (Czarkowski et al. 2011).

In the 2009/2010 season, Europe was dominated by influenza virus A, subtype A/H1N1/pdm09. Although many infections were caused by travelers entering a country, this virus prevailed in most countries in which it has already stayed and widely circulated. According to a WHO report, seasonal influenza viruses A/H1N1/, A/H3N2/, and type B continued to circulate, albeit at rather low levels. In Europe, the highest incidence and number of deaths due to infection with A/H1N1/pdm09 virus was recorded in the United Kingdom. In Poland, according to NIPH-NIH data, the number of reported cases of influenza infection was very low in the 2009/2010 season when compared with other seasons and also with data from other European countries. The peak incidence was exceptionally early in Poland as it took place in the last week of November 2010. Influenza A (H1N1)pdm09 was the predominant type A virus (MRC 2009). The 2010/2011 influenza season appeared 8–10 weeks later than that of the previous pandemic season in Europe, but still rather early compared to historical trends. In Poland, the peak incidence was recorded in the first week of February 2011 (Woźniak-Kosek et al. 2012). Influenza A/H1N1/pdm09 remained the predominant circulating virus, but in contrast to the previous season, there were a greater co-circulation of influenza type B and an appreciable downtrend in the percentage of influenza A/H3N2/ virus (MRC 2010).

In the 2011/2012 season, influenza virus activity began late in Europe. In Poland, the peak incidence was recorded on March 16–22, i.e., in the 11th week of 2012, and it was lower than those in the preceding seasons. In Europe, influenza A virus predominated and accounted for 90.5% of all infections, whereas influenza B virus accounted for 9.5%. A different level of influenza virus activity was recorded, depending on the region. In general, influenza A/H3N2/ predominated, but there was a discrepancy between the countries of Western Europe in which the virus was more than twofold more frequent than in Eastern Europe (MRC 2011). In Poland, due to a considerable percentage of unsubtyped A virus, it was hard to unambiguously confirm the predominance of A/H1N1/ pdm09. Nonetheless, the activity of influenza A viruses was confirmed in 21.4%. On the other side, circulation of influenza-like viruses was confirmed in 78.6% of all infections in Poland, with the predominance of RSV that accounted for 86.7% of influenza-like illness (Table 1).

In the 2012/2013 season, peak incidence was recorded in the 3rd week of January 2013. Influenza infections were recorded throughout Europe, with the predominating type A virus which accounted for 62% of all infections and clearly predominated over type B which accounted for merely 38% (MRC 2012). However, there were regional differences noted in the proportion of the two types as type A constituted 59% and type B 41% of infections in Western Europe whereas these proportions were 75% and 25%, respectively, in Eastern Europe. Concerning type A, subtype A/H1N1/pdm09 was recorded in 66%, and subtype A/H3N2/ in 34%. In Poland, influenza A/H1N1/pdm09 also predominated being responsible for 58.5% of all infections. In general, influenza virus activity was higher than that recorded in the previous season (Czarkowski et al. 2014).

According to a WHO report, type A virus predominated over the type B virus in Europe in the 2013/2014 season: 93% and 7% of infections, respectively. Concerning influenza type A, subtype A/H1N1/pdm09 was slightly more common than A/H3N2/: 56% and 44%, respectively. In the European Economic Area (EEA) countries, for a change, subtype H3N2 predominated over A/H1N1/pdm09 in a ratio of approximately 2.8 (MRC 2013). Overall, influenza viral activity was relatively very low and remained so even in the 8th week of 2014, when the highest number of influenza cases was detected. The predominating virus type differed among various countries. In terms of geographical spread, regional activity was largely limited to Western and Southern Europe. In some countries, detection of type A virus appreciably exceeded that of type B virus, and the proportion of subtype A/H1N1/ to A/H3N2/ demonstrated a substantial variability. In Poland, influenza A/H3N2/ virus was the predominant type A virus as it was recorded in 30.6% of all infections (Kondratiuk et al. 2016; Bednarska et al. 2015).

The peak incidence in the 2014/2015 season in Europe, including Poland, was recorded in the 8th week of 2015. For the entire European area, influenza type A virus predominance was confirmed in 69%, and type B virus in 31% (MRC 2014). Concerning type A virus, subtype A/H3N2/ (77%) predominated over A/H1N1/pdm09 (23%) (Bednarska et al. 2016c). The number of confirmations of A/H3N2/ significantly exceeded that of A/H1N1/pdm09 in most countries of Southern Europe. In Poland, due to a low number of samples tested of 2512, not all of which were sub-classified, influenza A/H1N1/pdm09 was the predominant type A virus (Bednarska et al. 2016d; Hallmann-Szelińska et al. 2016).

In the 2015/2016 season, peak influenza incidence was recorded in the 6th week of 2016 in Europe, while it was in the 7th week of 2016 in Poland (Szymański et al. 2017). Influenza type A virus predominated, confirmed in 70% of all infections in the continent, among which subtype A/H1N1/pdm09 was most frequent (49.7%). However, in Western and Southern European regions, influenza type B virus was sometimes more frequent (MRC 2015).

The 2016/2017 season was characterized by an early rise in infection transmission compared to previous seasons and a lack of clear pattern of influenza spread from west to east across the European area (Francis Crick Institute 2016). Influenza A/H3N2/ was the predominant type A virus, confirmed in 89% of cases. In Poland, type A virus also predominated (96.7%), but subtype A/H3N2/ was confirmed in just 29.7%. Due to a substantial percentage of unsubtyped material, A/H3N2/ predominance could not be conclusively confirmed (Szymański et al. 2018, 2019; Hallmann-Szelińska et al. 2018).

Influenza activity in the 2017/2018 season began in October 2017 in the European area. The season was also characterized by a relatively early rise in infection transmission that spread from west to east throughout the area (Francis Crick Institute 2017). Initially, influenza type B virus predominated during the season and most regions. The period of peak activity was somehow lengthened, comparably to the 2012/2013, 2014/2015, 2015/2016, and 2016/2017 seasons. The peak incidence of infections was reported in the 1st week of February 2018, amounting to 55%. In Poland, peak incidence was shifted forward to the 4th week of February 2018, but a significant rise in confirmations of influenza virus started already in the 2nd week of January 2018 and lasted till the 2nd week of March 2018. Influenza was confirmed in 96.7% of all infections, with type B/Yamagata lineage predominating in 65.5% of these infections. All European countries and regions with the ability to sub-classify influenza viral lines reported the predominance of Yamagata lineage in this season, whereas the frequency of type A subtypes varied (Francis Crick Institute 2018).

The peak incidence of influenza infections in Poland has been in January–March each season, since the pandemic influenza of 1968/1969 (Brydak 2008). In the epidemiological review above presented, the only exception from that pattern was the 2009/2010 season when the peak appeared in November of 2009. In nine out of the ten epidemic seasons, type A virus was the predominant source of influenza infection, whereas RSV predominated in influenza-like infections. There was an increase in confirmatory laboratory findings from season to season in relation to the number of biological samples examined, from 3.2 to 42.4% over the decade. Although the number of samples examined largely increased, it has remained just a fraction of a percentage point of the cases and suspected cases of infection, reported by health-care providers, which enormously increased from 568,958 in the 2008/2009 season to 5,337,997 in the 2017/2018 season. Influenza infection remains a health scourge and a major socioeconomic burden, with a dismally low yearly vaccination rate, which recently barely reaches about 3.5% of the general population in Poland.

Acknowledgments Funded by NIPH-NIH thematic subject 1/B.

Conflict of Interest The authors declare no conflicts of interest in relation to this article.

Ethical Approval All procedures and studies described in this review were conducted in accordance with the ethical standards of the institutional and/or national research committee and with the 1964 Helsinki declaration and its later amendments or comparable ethical standards. The article gained approval from the scientific board of the NIPH-NIH institutes in Warsaw, Poland.

Informed Consent As there was no current involvement of human studies in this review article, consent from individual participants was not required.

References

Bednarska K, Hallmann-Szelińska E, Kondratiuk K, Brydak LB (2015) Evaluation of the activity of influenza and influenza–like viruses in the epidemic season 2013/2014. Adv Exp Med Biol 857:1–7

Bednarska K, Hallmann-Szelińska E, Kondratiuk K, Brydak LB (2016a) Surveillance of influenza. Adv Hyg Exp Med 70:313–318

Bednarska K, Hallmann-Szelińska E, Kondratiuk K, Brydak LB (2016b) Innovations in the surveillance of influenza in Poland. Probl Hig Epidemiol 97:101–105

Bednarska K, Hallmann-Szelińska E, Kondratiuk K, Brydak LB (2016c) Antigenic drift of A/H3N2/ virus and circulation on influenza–like viruses during the 2014/2015 influenza season in Poland. Adv Exp Med Biol 905:33–38

Bednarska K, Hallmann-Szelińska E, Kondratiuk K, Rabczenko D (2016d) Molecular characteristics of influenza virus type B lineages circulating in Poland. Adv Exp Med Biol 910:1–8

Broberg E, Snacken R, Adlhoch C, Beauté J, Galinska M, Pereyaslov D, Brown C, Penttinen P, European Region WHO, The European Influenza Surveillance Network (2015) Start of the 2014/15 influenza season in Europe: drifted influenza A(H3N2) viruses circulate as dominant subtype. Euro Surveill 20(4):pii:21023

Brydak LB (2008) Surveillance of influenza. Influenza, pandemic flu myth or real threats? RYTM Publishing House, pp 165–190

Cieślak K, Kowalczyk D, Szymański K, Brydak LB (2017) The sentinel system as the main influenza surveillance tool. Adv Exp Med Biol 980:37–43

Cieślak K, Szymański K, Kowalczyk D, Hallmann-Szelińska E, Brydak LB (2018) Virological situation in Poland in the 2016/2017 epidemic season based on sentinel data. Adv Exp Med Biol 108:63–67

Czarkowski M, Romanowska M, Staszewska E, Stefańska I (2011) Influenza in Poland in 2009. Epidemiol Rev 65:199–203

Czarkowski MP, Hallmann-Szelińska E, Staszewska E, Bednarska K (2014) Influenza in Poland in 2011–2012 and 2012/2013 epidemic seasons. Epidemiol Rev 68:455–463

Francis Crick Institute (2016) Worldwide Influenza Centre WHO CC for Reference & Research on Influenza, London. Report prepared for the WHO annual consultation on the composition of influenza vaccine for the Northern Hemisphere 2016–2017. 22nd–24th February 2016. https://www.crick.ac.uk/sites/default/files/2018-07/crick_feb2016_vcm_report_to_post.pdf. Accessed on 15 Nov 2019

Francis Crick Institute (2017) Worldwide Influenza Centre WHO CC for Reference & Research on Influenza, London. Report prepared for the WHO annual consultation on the composition of influenza vaccine for the Northern Hemisphere 2017–2018. 25th–27th September 2017. https://www.crick.ac.uk/sites/default/files/2018-07/crick_nh_vcm_report_feb_2017_v2.pdf. Accessed on 15 Nov 2019

Francis Crick Institute (2018) Worldwide Influenza Centre WHO CC for Reference & Research on Influenza, London. Report prepared for the WHO annual consultation on the composition of influenza vaccine for the Northern Hemisphere 2018–2019. 19th–21st February 2018. https://www.crick.ac.uk/sites/default/files/2018-07/crick_feb2018_report_for_the_web.pdf. Accessed on 15 Nov 2019

Hallmann-Szelińska E, Bednarska K, Korczyńska M, Paradowska-Stankiewicz I (2016) Virological characteristics of the 2014/2015 influenza season based on molecular analysis of biological material derived from I–MOVE study. Adv Exp Med Biol 857:45–40

Hallmann-Szelińska E, Cieślak K, Szymański K, Kowalczyk D, Korczyńska MR, Paradowska-Stankiewicz I, Brydak LB (2018) Detection of influenza in the epidemic season 2016/2017 based on I-MOVE+ project. Adv Exp Med Biol 1114:77–82

Kondratiuk K, Czarkowski M, Hallmann-Szelińska E, Staszewska E, Bednarska K, Cielebąk E, Brydak LB (2016) Influenza in Poland in 2013 and 2013/2014 epidemic season. Przegl Epidemiol 70:407–419

MRC (2008) National Institute for Medical Research. WHO Influenza Centre, London. Report March 2008. https://www.crick.ac.uk/sites/default/files/2018-07/interim_report_mar_2008.pdf. Accessed on 15 Nov 2019

MRC (2009) National Institute for Medical Research, WHO Influenza Centre, London. Report February 2009. https://www.crick.ac.uk/sites/default/files/2018-07/interim_report_feb_2009.pdf. Accessed on 15 Nov 2019

MRC (2010) National Institute for Medical Research, WHO Influenza Centre, London. Report prepared for the WHO annual consultation on the composition of influenza vaccine for the Northern Hemisphere. 14th–18th February 2010. https://www.crick.ac.uk/sites/default/files/2018-07/interim_report_feb_2010.pdf. Accessed on 15 Nov 2019

MRC (2011) National Institute for Medical Research, WHO Influenza Centre, London. Report prepared for the WHO annual consultation on the composition of influenza vaccine for the Northern Hemisphere. 14th–17th February 2011; https://www.crick.ac.uk/sites/default/files/2018-07/interim-report-feb-2011.pdf. Accessed on 15 Nov 2019

MRC (2012) National Institute for Medical Research, WHO Influenza Centre, London. Report prepared for the WHO annual consultation on the composition of influenza vaccine for the Northern Hemisphere. 14th–17th February 2011. https://www.crick.ac.uk/sites/default/files/2018-07/interim-report-feb-2012.pdf; https://www.who.int/influenza/vaccines/virus/recommendations/consultation201209/en/. Accessed on 15 Nov 2019

MRC (2013) National Institute for Medical Research, WHO Influenza Centre, London. Report prepared for the WHO annual consultation on the composition of influenza vaccine for the Northern Hemisphere 2013/14. 18th–20th February 2013. https://www.crick.ac.uk/sites/default/files/2018-07/interim_report_february_2013.pdf. Accessed on 15 Nov 2019

MRC (2014) National Institute for Medical Research, WHO Influenza Centre, London. Report prepared for the WHO annual consultation on the composition of influenza vaccine for the Northern Hemisphere 2014/15. 17th–19th February 2014. https://www.crick.ac.uk/sites/default/files/2018-07/nimr-report-feb2014-web.pdf. Accessed on 15 Nov 2019

MRC (2015) National Institute for Medical Research. WHO Influenza Centre, London. Report prepared for the WHO annual consultation on the composition of influenza vaccine for the Northern Hemisphere 2015/16. 23rd–25th February 2015. https://www.crick.ac.uk/sites/default/files/2018-07/nimr-report-feb2015-web.pdf. Accessed on 15 Nov 2019

Szymański K, Kowalczyk D, Cieślak K, Brydak LB (2017) Regional diversification of influenza activity in different in Poland during the epidemic 2015/2016 epidemic season. Adv Exp Med Biol 1020:1–8

Szymański K, Kowalczyk D, Cieślak K, Hallmann-Szelińska E, Brydak LB (2018, 1108) Infections with influenza A/H3N2/ subtype in Poland in the 2016/2017 epidemic season. Adv Exp Med Biol:93–98

Szymański K, Cieślak K, Kowalczyk D, Hallmann-Szelińska E, Brydak LB (2019) Respiratory viruses in different provinces of Poland during the epidemic season 2016/2017. Adv Exp Med Biol:1150, 83–1188

WHO (2019) The Global Influenza Strategy 2019–2030. https://www.who.int/influenza/global_influenza_strategy_2019_2030/en/. Accessed on 14 Nov 2019

Woźniak-Kosek A, Czarkowski MP, Staszewski E, Kondej B (2012) Influenza in Poland in 2010. Epidemiol Rev 66:599–604

Printed in the United States
By Bookmasters